HOLDING ON and Letting Go

a life in motion

AF704512

LINDSAY SWOBODA

Elva Resa ✳ Saint Paul

Holding On and Letting Go: A Life in Motion

This is a memoir, written in creative nonfiction style (a true story, supported by elements of fiction), reflecting personal thoughts and happenings as remembered by the author. Some names, events, and dialogue may be changed, compressed, or recreated. The author says, "I have told my story as faithfully as I can, using memories and written records. I have an excellent memory for detail and interviewed other people involved in this story when possible to be true to the events I recount. Also, I keep copious notes, proven by the journals that account for a considerable amount of our household goods weight allowance."

A portion of the author's story has appeared in previously published essays: "The Question, The Answer," *The Line Literary Review*; "Vacation Pie," *Legacy Magazine*, Vol. VII; "Rooted in Hope," *Legacy Magazine*, Vol. VIII; and "Settling in Again," *InDependent*.

Cover painting and handwritten font by Lindsay Wilkins for Elva Resa
Senior editor: Terri Barnes
Senior copyeditor: Christy Lyon

Library of Congress Control Number: 2025933242
ISBNs 978-1-934617-86-1 (pb), 978-1-934617-85-4 (ebk)

Printed in the United States of America.
1 2 3 4 5

Elva Resa Publishing
8362 Tamarack Vlg, Ste 119-106
St Paul, MN 55125

ElvaResa.com
MilitaryFamilyBooks.com

To Team Swo:
Ryan, Evelyn, and Hunter

We never give up—
on each other or ourselves.

A WORD FROM THE AUTHOR

No Getting Around It

AS A TODDLER, MY SON BECAME MILDLY OBSESSED with the book *We're Going on a Bear Hunt* by Michael Rosen. In the story, a family decides to go on an adventure in search of a bear, but they bump into obstacles along the way, such as a dark forest, a muddy patch, tall wavy grass, and a deep cave. Each time they approach a hurdle in their path, they repeat the lines: "We can't go over it. We can't go under it. Oh no! We've got to go through it."

On my 125th reading (but who's counting?), this phrase finally sank in. While it is tempting to find a prescriptive set of steps to help us get over or under an obstacle, ultimately we've got to go through it. And even when we've gone through our trials, we still may not understand why or what they mean without the perspective of time and heartfelt unpacking with a trusted confidant or counselor.

I know. *Sheesh*. I just made a life lesson out of a fun book for kids. But what if we arrived at the other side of this story more present, whole, and with great hope? What if we built a remarkable and meaningful life along the way?

As my family both endured and enjoyed military life, we met frequent upheaval, exhausting unknown challenges, and invitations to pause, pursue growth and intense connection, and forge new dreams for ourselves. I discovered one constant we could count on was change, and while we grew in our capacity to navigate what we were called to next, each chapter asked us to stretch.

Sometimes that stretching felt nourishing, like a healthy yoga session, where our bodies and minds adjusted and felt stronger afterward. Other times it felt like we were being stretched to a point of snapping, like the sickening pop of a twisted ankle.

To make sense of it all, I often turned to the pages of my journal and used its safe space to unleash my stories in an effort to process them. I held them up to the light with my pen, uncovering a spectrum of experiences in shimmering ink. A dear mentor of mine once told me that to find the essence of her story, she had to write through it, to put words on paper to both craft an understanding of challenging times and to unveil beauty that might have been there all along.

Over the past nine years, I have begun to open my journal and share my writing with fellow travelers. The journey to move my story into a book has been as full of joy and fulfillment as the life it chronicles. It has also been punctuated by days of questioning, unforeseen challenges, and self-doubt. I have discovered I can choose to notice the great and small joys before me and to also name my losses and grieve them. I have learned there is a balance in this life between holding on and letting go.

I've often wondered whether I can convey this story of mine, this balance of joy and sorrow, and if I can, will it make a difference? Conversely, what would happen if I kept my story safely nestled in my own pages, seen only by me? What richness might I miss by choosing safety and silence? Choosing to write this book has been a way for me to celebrate the life we have lived and the one we are going through day by day.

One day, when my daughter was eight years old, we were working together on a family journaling project. While she painted, I talked to her about the journal prompts.

"Okay, next think of a character from a book or movie and tell me about a time when the character believed in themselves," I said.

"Does it have to be from a book or movie?" she asked, glancing at me as she spread deep blue paint across the top of her page.

"No, I don't think so. What did you have in mind?" I asked.

Evelyn looked at me and said, "My mom."

"What?" I asked, meeting her eyes.

"Are you going to write it down?" she challenged. I picked up the pen and wrote as she narrated it to me.

"When my mom turned in her book she had to be brave—because Mommy thought she wouldn't be good enough. But we knew she was."

I stared down at the page and allowed her words to sink into my soul. Evelyn patted my hand and went back to her painting.

Sharing stories, whether we are the readers or the writers of those tales, helps fuel our family with the perseverance to keep going with renewed intention. We fan the flame of hope. We open our arms as a comfort and to be comforted. We hold what we can, and we let go of what we must. And when we pursue the story of our hearts, we grow curious. Like a child gently turning over stones in a garden, we discover what lies beneath, be it a tangle of roots, a nest of pill bugs, or the flash of crystal quartz. We bring our narrative into the light.

We grow in our knowledge that, ultimately, our moments of despair do not define us. Instead, they are obstacles that make us stronger when we go through them.

PROLOGUE
Wish Upon a Star

"I DO BELIEVE IN FAIRIES! I DO, I DO! I do believe in fairies! I do, I do!" My daughter and I chant over and over again with Peter Pan as he stands protectively over Tinker Bell's apparently lifeless body in the Disney movie we are watching. Evelyn's long hair, still damp from the shower, sticks to my neck as she leans her head back onto me. It smells like strawberries.

We are shouting at the one wall in our house that is devoid of photos or decor. When we moved into this house, I fought Ryan over his insistence on leaving this wall empty. The immensity of sharp whiteness is bleak compared to the rest of our home design. But in his quiet, steady way he convinced me, and he was right. This empty wall has become life-giving. Once a week, its pure blank whiteness becomes a portal to adventure, a silver screen for our family movie night, cascading with Moana's expansive ocean, swirling with snowflakes around Anna and Elsa. Tonight, it is whisking us off to Neverland.

"I do believe in fairies! I do, I DO!" we cry out one last time, and the mighty pixie begins to glow. She sparkles and sputters back to life and my daughter's smile gleams. As the pirate ship sails high above London and into the night, I can feel her trembling energy. Tears brim in my own eyes. Whether making an actual trip across the world or an imaginary one in our living room, I will forever delight in soaking up every minute of our adventures together.

After the movie, we battle with pillows, and Ryan lifts Evelyn high as she leaps from the couch and flies through the air. We brush our teeth and gather to sit on her bed, looking out the window before lowering the shade. Staring into the night sky, we count the lights still blinking on the mountainside.

"Mama, I want to fly like Tinker Bell. It is my greatest wish." Evelyn's earnest eyes search mine, and my heart simultaneously breaks and smiles.

"Did you know that every night when I was little, I stood by my window in my childhood home and picked the very brightest star and gave it a wish? It's funny because I had a similar wish—I wished I could fly without wings. Would you like to try a wish?"

Urgently, she nods and points out several blinking satellites and one airplane before we settle on a true twinkling star. I teach her the magic words:

Star light, star bright
First star I see tonight
I wish I may, I wish I might
Have this wish I wish tonight.

She sends her wish up to the star with her hand—wet with the whisper of a kiss.

"I wish I had wings to fly like a fairy."

We rest our heads with the warm blossoming of hope in our hearts that night. It is magic to be together under this one roof. We welcome happy moments and notice these tiny treasures: stories on a white screen wall and popcorn kernels clinging to the carpet. Our dreams fill the darkness with pirates, lost boys, mermaids, potential new leaps to foreign lands, and three heads on pillows.

In the morning, I wake to fierce cries from her bedroom. It is a different sort of sob than I've heard in the past. It does not sound hungry or tired. The wailing sounds scared and frustrated. Rushing in, I see her holding Bunny tightly to her chest, staring up at the ceiling while hot tears pour down her cheeks.

"I wished, Mama, I wished, and I don't have wings! I woke up, and there are no wings!"

Her three-year-old face is red and puffy, her brows knit together to form a tight crease on her forehead. Her anger with me permeates the air between us. I promised the magic of stars and did not deliver. My brain searches for the proper parental response to the conundrum I created by getting carried away with childhood magic. The silence stretches between us as I sit down next to her and lightly brush my fingertips through her hair.

"Did you ever get your wish, Mama?" her lip pokes out into a pout; her eyebrows raise in a challenge.

"You know what? I did, in a way. I might not be able to soar through the air like Superman or flap and glide like a bird, but my wish was granted in other ways."

I have piqued her interest. She closes the gap between us on the bed, resting her head on my thigh. Maybe there is hope of saving some magic today.

"I have flown all over the world," I tell her. I remind her of the kind host family I lived with in Italy and the cruise ship where I once danced. I stroke her hair while I share, "Daddy and I have crisscrossed the whole globe together, taking chances on places and people. When we had you—"

She turns her head in my lap and gazes up with a smile.

"When we had you, Daddy had to leave us to work for a while, so Mommy took you everywhere! We were travel girls, just you and me, singing and swinging on planes."

Her eyes are soft again, connecting the dots, and I feel her reassurance swelling, her soft brown hair like fine silk on my fingertips.

"I got my wish and more," I say. "I needed airplanes to fly, but I have flown everywhere."

"And we flew here, Mama, to our new home. I like to fly."

"Me too. We'll be flying again real soon to see our family. Now," I say, tickling her tummy, "are you ready for some breakfast?"

Our day continues with the normal routine: breakfast and reading in our overstuffed yellow chair. We clean house, walk to the park, and slip into the magical mundane of everyday living. That's the thing about being a military family and expatriates: We have moved and traveled, planted our feet all over the globe, and yet there are always more days of peanut butter and jelly sandwiches, laundry, and living room play-tents than great sweeping journeys.

I do still believe in magic. I do, I do. There is grand magic in leaping into the unexpected and taking a chance on both myself and the mysteries of the world. There is magic in waiting for what I want, perhaps even magic in getting what I need instead. Magic lingers in both camel rides across the desert and in the sweet squeeze of your toddler's hand during a bedtime story.

Ryan and I have climbed aboard the wrinkly backs of elephants in Thailand and equally loved a good old-fashioned hayride in Ohio. We are lifted by the pixie-dust wonder of how we view our journeys and ourselves. We already live extraordinary lives—but we have to choose to see how our ordinary morphs into magic.

I used to think my success was measured in work accomplished, passport stamps gathered, exotic adventures logged. But now I want to see the accomplishment in each day's journey. I want to acknowledge the value of quiet, soft stories as well as the loud, vibrant ones. I have tales to tell about castles in Europe and stories of how my eighty-six-year-old nana's eyes dance when she sees me walk the familiar path up to her door.

While it's hard to imagine my passion for traveling will diminish—and I hope my suitcases never remain unpacked for long—I am learning to be just as happy at home with a string of plain, ordinary days.

Weeks later Ryan and I jaunt up our sidewalk for a night stroll, our daughter swinging between our arms. She gazes up breathlessly into the sky and asks me to teach her the star wish again. She has not asked since that disappointing first attempt. I call it up into the inky blackness, and she wishes for wings again. Her face is bright and hopeful.

I pick my own star and make a silent wish.

> *Let her realize her wings can come true in multiple ways. Give her courage to let her dreams change and grow along the way. Let her come alive in her own story. Let her always be able to come home to herself.*
>
> *Let her see me noticing my extraordinarily ordinary life. Let me be brave enough to keep changing too.*

I walk up the road with my small family, soaking in our current stillness, primed to answer the call of adventure when it next dials in. We will keep our wands at the ready and allow wishes on stars to be made at late-night windows. I will allow myself to not only search for magic but to be lost in it.

A life well lived is one where we notice the life we are living.

Choosing My Path

IT IS A COOL MORNING IN BOWLING GREEN. The trees that line College Street have the first flicker of autumn on their tips. A morning chime sings from the top of Cherry Hall on the Western Kentucky University campus. I quietly close the door of my apartment and adjust my backpack on my shoulders. The door is red. It reminds me of Hester Prynne's scarlet letter and the way her circumstances pulled her apart at the seams. I'm feeling pulled apart too, though my circumstances are very different. Maybe they are not the problem. Maybe I am.

I start making my way up the sidewalk to my first class. Ballet at nine fifteen in the morning is brutal but the pain of working my body distracts my brain. A leaf crunches under my foot and I think the sound must be similar to the cracking of a heart. It has been a year since I graduated high school, all smiles in my pressed crimson gown. I threw my cap into the air. I collected my diploma. I came to college with a plan. But now it's my sophomore year and I still feel like a sprouting seed just beneath the soil. Green and round, a small sprig that doesn't even know what color it will be, what its leaves will look like, or the size and shape of its petals. I want the freedom to find out, but I'm terrified to find out alone.

I think of him. Ryan was there at my graduation party, handsome in a blue shirt, his cuffs rolled up to reveal the fuzzy part of his forearms. I remember pressing my forehead up to his. I traced the familiar lines on his hands with my fingertips and breathed in the blend of his spicy cologne, pine needles, woodsmoke, and warm sunshine.

We didn't live in the same town, but we met in my junior year thanks to my friend Rachel, who invited me to join her family on a cruise. My parents agreed I could go as long as I paid my own way. I scraped the inside of the popcorn machine for months, saving the minimum-wage paychecks I earned at the dollar theater.

I remember the excitement of departure. Rachel and I rolled our luggage up the gangplank and into our small cabin. Stampeding feet thudded down the hallway, and we peered out to see a herd of teenage boys gathered only a few doors down from us. We slammed the door shut with giggles and slipped into more appropriate cruise wear: halter tops, shorts, and sparkly flip-flops.

We met the boys out in the hall and were surprised to hear they were from Ohio as well. We made small talk until we hit an awkward pause, and Rachel and I turned to go.

"Hey—uh—what kind of camera is that?" one of the boys asked, in an obvious delaying tactic. Grabbing my wrist, he slid the camera into his hand.

"What kind of vacation have you ladies been having already?" He chuckled and turned to scroll through my photos on the view screen.

My cheeks flushed bright red. "That's enough! You can give it back now," I said, straightening my shoulders and back. I tried to grab the camera, but he laughed and dodged. My hand sliced through only air, and he ran down the hallway into his room.

I shrugged, raised an eyebrow, and curled my lips into a tight smile.

"You boys tell him I'll be back." Linking arms with Rachel, we flip-flopped away as fast as we could.

Later that night in our room, we heard the boys pass by the door. I knew it was time to make good on my promise. I went out the door and down the hallway, Rachel at my heels. I knocked. This time, I would not leave without my camera.

The door opened, but instead of the camera thief, a different boy answered. He had curly brown hair, honey eyes, and only dress pants on. Not wanting my eyes to drift to his bare chest, I looked down at his feet. They were also bare, and if toes could be attractive, his were.

"Oh-hi! Sorry, I thought you were one of the moms," he said, grinning.

"One of the guys in your room took my camera earlier. I need it back."

"One sec," he closed the door. I heard muffled voices, and soon he returned with the camera. As he handed it over, his fingertips grazed mine.

"I'm Ryan. See you around," he said with another flash of a grin.

"See you," I said as I turned and walked to dinner with Rachel.

Over the next few days on board, we bumped into the boys accidentally-on-purpose until it was established we were all officially going to hang out. We played cards together in the hallway. There was pizza at the midnight buffet. Ryan told me about his parents' large woodland property and his rottweiler, Gabby. At one of the evening dance shows, the boys sat next to us.

On the last night of the cruise, Ryan and I met under the stairs. We talked as the boat gently rocked us back to port, and we decided to exchange numbers. A few months later we were boyfriend and girlfriend, driving the hour and a half between our houses to meet up for dates, grinning together in homecoming and prom pictures. A year of long-distance calls on our bedroom cordless phones, talking for hours about our hopes and our dreams.

His face, his smell, the honey eyes, and the way he smiled at me felt like things I had always known. Like things I would always know.

He enlisted in the Marine Corps the same year I entered college. We made it through one year apart, but then we crumbled. Or I crumbled. I thought being high school sweethearts would last forever—that there was some powerful magic in first love.

I look up at the path ahead leading me to my first class, then down at the sidewalk at the leaf I have crunched beneath my foot. I spread its orange pieces out with the toe of my shoe. This leaf has completed its season of life. It was meant to fall. Another leaf, pale green and not quite ready to let go of the branch, is ripped loose by the wind. It ripples down and my jaw clenches.

I had it all just a few months ago—the job, the high school sweetheart, the perfect plan for my college career. Now what? I've smashed it into a million pieces. The brisk air makes me shudder. I breathe on my fingers to warm them and pull my phone from my pocket. I scroll to Mom's number. I need to hear her voice. It rings and connects. I can hear busy office chatter and telephones. She is already at work.

"Lindsay?" she asks.

"Hi, Mom." I pause and watch a red leaf plummet down from the trees. I want to tell her I don't know what I'm doing, but instead I say, "I'm walking to class."

She waits on the other end for me to continue. I can hear her fingers clicking on her computer. I hear the sound of a file drawer sliding open and shut while I search for words to explain how I'm feeling. A phone rings.

"Hold on, honey," she says and goes to answer it. "Okay, I'm back. How are you?"

"Oh, I'm okay." My courage to confess has seeped out with every step up the hill. So I tell her about a movie I watched, my homework, and how I'm anxious to see the casting notices for the December show. These arbitrary details wedge perfectly into the space between us.

"I made it to class," I say, pressing the phone into my cheek a little harder as I jog across the street.

"It was good to hear from you, hon," Mom says. She pauses, and I can feel her pressing her face closer to mine from all these miles away. She sucks in a breath and exhales, "You will be okay, right?"

I nod as if she can see me. She knows there's more to say, and I know it too. But I'm not ready to tell yet. "Yes, of course."

We hang up as I approach the door to the dance building. Several of my classmates are out hanging around on the steps. Puffs of cigarette smoke swirl into the air, and one dancer chugs a Coke. My stomach grumbles. I sling my backpack down at my feet and pull out a granola bar. It is gone in a few bites. I lick the last crumbs from my wrapper and head upstairs to the studio. As I enter and take my place at the ballet barre, I look up at one of the many motivational signs our director has placed above the mirrors.

The bright red letters urge: "Don't TRY. Do." Some of my classmates scoff at these flimsy paper reminders, but the words are emblazoned on my mind.

After class, we gather in the hall, angling for position to see the casting notices posted on the wall for the December show. My friend Steve and I see we are in nearly every number together.

"Well, our lives are over," he grins. Rehearsals are four evenings a week for two hours. Given the number of castings we have received, we will be here every day until the December show.

"Want to grab lunch?" Steve asks.

"Sure." We walk down through the building arm in arm.

"How's Ryan? Do you know?" I look at him sideways and he stares straight ahead, pinching my arm in the crook of his elbow. I am surprised to feel anger at his question. I know he asks because he cares, but I do not want him to care. I don't want to care.

"I don't actually know," I say, challenging him with a direct stare. He shrugs and we push open the doors to the cafeteria and order our food. We talk about what the choreography and music might be like and dream a little bit about what the costume designs might be. Our conversation about the show lifts me again. My love life might be a mess but who cares? That's not the path I'm on right now.

It's the work I'm after. I need to walk the path I've chosen, and all will be well.

Times Square Shimmer

THE REST OF THE SEMESTER FLIES BY in a flurry of rehearsals and ends with the final dance performances. Once they wrap, we have the final round of auditions for the spring musical, which is *Ragtime*. The day of the auditions, I am jittery as I enter the dance call. The guest director is here for casting. From New York City. Everyone in the department is humming with excitement. Getting a part in this musical will be a stepping stone toward worthiness. Making an impression on the professional director would be another. I enter the studio in my leotard, tights, and heels, hair pressed into a neat low bun and my makeup immaculate.

The black marble floor is packed with hopeful dancers. I catch glimpses of everyone's eyes in the mirror. They are hungry like me. Starving for recognition. Auditioning is always like this. Screaming in the middle of a crowded room.

After the audition, I am smiling. In spite of the nerves, I enjoyed it. After years of training, this is my why—because I feel like myself on the floor. I belong out here.

The turnaround for the *Ragtime* cast list is quick. It is up within a few days, posted in the hallway. I check it before my ballet final and heave a sigh of relief as I hit my number. I am in.

From his office, the dance director motions to me.

"Excuse me, Lindsay? I need to see you."

I sit across from him, and he tells me the New York director who came to the audition was very impressed with me. Five other girls from the department and I are invited to apply for three open spots in a Christmas show—in New York.

"You need to bring me your headshot and resume later today," says the dance director. "Good luck," he adds in dismissal.

He does not ask if this is what I want to do with my Christmas holiday. He does not check to see if I would rather spend it with my family back home. He knows every dancer is desperate for a chance like this. I am desperate for this chance.

As a child, I made a dance bucket list and stuffed it under my mattress. "Dance in a show in New York" was high on the list.

Even then, I was wishing for wings.

On my way to my dorm, I make a phone call. The office chatter fills my ears again as I explain the situation to Mom.

"It's New York, Mom!" I say. "I have to do this if I get it. I have to," I tell her, but my hand clenches into a fist waiting for her response. I have never missed a Christmas with my family.

"Well it all sounds exciting," she says, but her voice does not sound bright like it normally does. "If it happens, we can always celebrate Christmas late. We'll keep the tree up for you," she plans aloud.

"Really? That would be great, Mom." I smile and wait, but she does not offer more. We say goodbye, and I wonder for a moment if my parents are still growing up too. If we are all still trying to figure it out.

A few days later, I find out I am one of the three dancers chosen. This is it! The girl from suburban Ohio, going to dance in a New York show at Christmastime! Every niggling doubt about my success washes away while I pack.

~

I arrive at the airport on Long Island late on Friday night and meet up with Sarah and Anna, the two other dancers. Wreaths and red velvet bows bring holiday cheer to the bustling space. I catch a glimpse of myself in the airport bathroom mirror: a wide-eyed girl eager for possibility.

The producer of the show picks us up and takes us to the large white mansion where some of the cast are housed. It's impressive from the outside—surrounded by trim hedges and a black wrought-iron fence—but freezing inside, as we soon discover. Anna and I share a room at the very front of the house and the street light beams through the thin white curtains.

"Let's put some of our blankets over the window. It will seal in the warmth and block out the lights," Anna says. We balance on our tiptoes on her bed to cover the window. The temperature doesn't chill our enthusiasm for this adventure.

"Can you believe this?" Anna says. Her face beams.

"No. This is crazy!" I say.

"Are you nervous?" she asks and sits down on her bed to kick her boots off.

"Yes," I admit and start to pull off my jacket, but the chill stops me. Instead, we put on more layers of clothes and use our coats as extra blankets. It seems I will never fall asleep, yet I wake to voices outside our door. It is morning, and my dream is coming true.

The first day is a blur of greeting the other cast members and stuffing choreography into our brains and bodies. It's Saturday, and we have today and tomorrow to learn the entire show. We have Monday off, then we'll be out of the practice room and on stage for technical rehearsal, adding lights, costumes, scenery, and the orchestra. Sunday will be the first show. The schedule makes my head spin but my determination to prove myself out-weighs all fear.

The next day, Sarah, Anna, and I bundle into our coats to do some grocery shopping. The air is crisp and the sky is bright blue. Our boots crunch through snow patches on the way. We follow two other dancers. One has curly red hair that bounces on top of her head and the other has shiny brown hair that sweeps over her shoulder as she turns back to look at us.

"So, how old are you?" she asks.

"Nineteen," I answer. "What about you?"

"I'm twenty-seven."

"Thirty-two," says the other.

As we walk, they talk about their previous jobs. One of them has worked for several cruise line companies and tours. Big names roll off their tongues, from Broadway tours to international tours across Asia. I absorb every detail and wonder how she unlocked all of these incredible contracts. The other woman used to be one of the Radio City Rockettes, but a hip injury landed her at this Christmas show on Long Island instead of in Manhattan.

They talk about MetroCards and subletting apartments from Manhattan to Astoria and into Brooklyn. In their downtime between contracts they wait tables or do other odd jobs.

"You just do what you have to, ya know?" The brunette turns to open the door of the store.

Back at the house we stash our groceries, eat a quick snack, pull on our heels and leg warmers, and head to rehearsal. We run through the entire show from top to end and I delight in my hands and legs moving in unison with everyone else. In one number I am next to the red-haired dancer, and it feels even more special to kick next to a former Rockette.

At dinner, I slurp ramen and the steam warms my face as I giggle at another joke from Sarah. The glow of the kitchen is warm and soothing. I'm doing it, I think. I always wanted to dance in a show in New York, and while I have only made it to the outskirts of Long Island, it feels like this dream is on its way to realization. The two dancers from the grocery trip come into the kitchen to kiss cheeks and dole out goodbye hugs.

"Where are you going?" I ask.

"Back to the city," says one. "I've got a shift tomorrow."

"Me too," says the other, pulling her shiny hair back into a ponytail, then crushing it under a fuzzy hat. "Gotta pay that rent."

We discover that most of the dancers in the house are leaving in the morning. There's a train station not far from the house. We can be in New York City in approximately one hour and forty-five minutes. "Do you want to go out tomorrow?" Sarah asks.

"Heck yes," Anna says.

"I'm in if you are," I say.

As we step off the train at Penn Station the next day, we join a mass of people flowing like ocean waves past the train. When we are amongst them, they buffet the sleeves and back of my thin black pea coat and roll across my slow-moving feet. Anna grabs my hand and we surge against the tide, breaking through to the front doors where we spill out onto the black sidewalk speckled with gum. Instead of seaside gulls crying overhead, I can hear the coo of pigeons and honks from taxis. We head straight for Times Square. Anna and Sarah laugh as they watch my reaction; they've both been to the city before, but my eyes widen with each step.

We dodge traffic and land right in the middle of the square, where I spin around and open my arms to the neon billboards.

In my head, I'm bursting into a song from the musical *Annie*:

Oh, NYC! You make 'em all postcards.
You crowd, you cramp. You're still the champ!
Amen For NYC. The shimmer of Times Square!
The pulse, the beat. The drive ... NYC!

I pull out my camera. Anna snaps several photos of me smiling so big it hurts. We get last-minute tickets to a matinee at Radio City Music Hall. After the show we walk to Rockefeller Center and gaze at the big, beautiful Christmas tree and ice skaters stumbling about on the rink. It is easy to play tourist in this town.

The wind bites at our cheeks as we walk. We give ourselves breaks from it by ducking low under the pavement and into the subway. I think of the older women in the cast with their metro tickets, and I buy my very own. It is a flimsy piece of cardstock, but this one little rectangle gives me access to the underground network, a pass to the entire city. Our next stop is a walk through Central Park, where we discover Strawberry Fields and the *Imagine* mosaic, which is covered in flowers.

"My dad would love this," I say and crouch down next to the circle. Anna snaps a photo, and my smile falters. We walk on and Beatles songs play in loops in my mind. I feel the first pang of missing my family. I can see my dad in his winter sweater and thick wool socks perched on a stool in the living room playing along to "Blackbird" with his guitar. I always hum the lyrics with him.

All your life, you were only waiting
For this moment to be free

Continuing through the park, we pose with statues and breathe in fresh air. It is a wonder to go from being surrounded by gray concrete to gazing up at giant trees. Many of them are stark and leafless. Their limbs reach up to the thin winter sunlight like twisted bones.

They feel familiar. I thought I would love the incessant hustle and bustle of the city, but in the open air of the park and the stirring of trees, I feel a whisper of home. I run my hand against a fluffy fir tree. I look up at the top and watch it shimmer in the afternoon light. It is more beautiful than every billboard in Times Square.

Our day ends with a final stop. It has been four years since the 9/11 attack, and as we approach the site of the former Twin Towers of the World Trade Center I am struck by the emptiness. It is a large construction site now with the promise of a memorial to come. I press my palms to the chain link fence surrounding the dark, deep hole and peer down into the void. Flowers and photos, notes, and other evidence of precious memories flap against the cold metal.

I was a sophomore in high school when it happened. The TV in the room across from ours was on, and we crowded around the door of our classroom as the second plane hit. The lockdown bell sounded minutes after and I sat against the cold white brick wall with my classmates. The teacher turned the lights off and we sat with our heads in our hands. We waited. School was dismissed early, and all students were told to go straight home. My sister found me in the hallway and we held hands out of the school doors. We cut across the public park and ran the last two blocks home. Mom, a school nurse, and Dad, an elementary school principal, were still at work, trying to help kids find parents and safety. Inside our house, my sister and I cried and waited.

It was the end of a piece of my childhood innocence and the rupture in my belief in happy endings. There is not a reason for everything. Not everything turns out well.

Ribbons flutter from the chain link fence and I think of all the tears here too. I think of the end of lives and dreams. I think of Ryan in his uniform at boot camp graduation. I think it is only a matter of time before he heads out to stand in a war.

Just last year, I surprised him at his graduation from boot camp on Parris Island. I took all my exams early to get on my first plane ride to watch him stand in excellence. I remember how we watched the platoon march down the sidewalk.

His mom kept saying, "There he is! See him? There he is!"

"Oh yes! There he is!" I replied. In reality, I couldn't find him.

I couldn't pick my Ryan out of all the sea of marching men and women. When I hugged him that day he was both himself—and not. I was both me—and some other girl-woman entirely.

I grip the chain link surrounding Ground Zero and wonder what Christmas will be like this year without him.

Our train ride is solemn, heading back to the cold cast house and endless rehearsals. I nod off against the cool glass window. In dreams, the shimmer of Times Square swirls against black holes.

The rest of our week whirls in a flurry of costume fittings and rehearsals. The night of the first performance I stand waiting for my entrance. I can hear the hum of the live audience on the other side of the curtain. The excitement of having all eyes on us, hearing real laughter and applause, is irresistible. We are drawn like moths to the flame.

I allow myself to soak up the light of the world we have created. There is strength and confidence when the singers cascade through their songs. A dancer glides seamlessly through a tap solo. The backstage crew lifts and turns the sets in fluid waves of action. In the final scene, I swirl onto the stage in my green velvet dress. Fake snow falls down upon us as we clap, sing, and bow.

After the show everyone hustles into jeans, sweaters, coats, and boots. We pile into the company van and head to a nearby restaurant for a round of celebratory drinks and chicken wings. Cast members eat, drink, and laugh together. We college dancers seem to have passed the test. We learned all the material. We performed a successful show. Anna clinks her mug with me, but when we are alone, she looks down at the table.

"Do you think this is all it is?" she asks.

I raise my eyebrows, knowing what she's getting at.

"All it ever will be? I don't know."

The shimmer is dimming. I thought I knew what a theater life would be. In my childhood bedroom in Ohio, this dream lit up my mind like twinkling diamonds.

I felt joy on stage tonight but I also felt a pit. And I don't see a ladder leading up. What is the pinnacle of this job? It feels like the peak is working in the theater. Between jobs it is back to audition rooms and side hustles.

Back at the cast house I flop on my bed. Anna and I whisper to one another across the quiet room about job security, retirement, the Actors' Equity union, and whether or not we have what it takes. Our late-night food and drink have led to philosophy. We speak with the truth serum of mild inebriation. We hover over whether we love it enough.

I roll over in my bed and squeeze my eyes shut, but the thoughts keep coming.

What have I found here? I have loved this work and the idea of this work, but it is starting to dawn on me that the work does not love me back. The work may never love me back.

Why is it easier to love the work more than myself and my relationships? What does the other choice look like?

~

Penn Station spews Anna and me out to the sidewalk again. This time we have a map to Hell's Kitchen and backpacks pulling at our shoulders. The show is closed from December 23 to 26, so we have time to ourselves. Sarah is headed back to Kentucky for the holiday, but Anna and I are staying put in New York. We're thankful we do not have to sit in the empty cast house by ourselves. Our stage manager generously offered the use of her apartment, conveniently located a few blocks from Times Square. We'll have everything we need for a glamorous NYC Christmas.

Christmas Eve morning, the sun glistens off the water and taxis honk below. For the first time in my life I won't be at Nana and Grandad's house for our family's traditional Christmas feast and celebration.

I let a few tears slip out before pulling on my boots and linking arms with Anna to head out into the city again. We want to see a show, and fortunately, most productions on Broadway are still open. But when we check times and prices, we can't afford any of them while still keeping enough money for our food for the next few days. We end up buying tickets to *Beauty and the Beast* for twenty-five bucks each from a man on the street with a likely story. With forty minutes to kill before the doors open and we discover whether the tickets he sold us are legit, we drift in and out of Broadway knickknack stores.

I see a *Phantom of the Opera* T-shirt, the mask and red rose design gleaming on black cotton. A showing of *Phantom* at our downtown theater was the first date I took Ryan on with my own money. I saved up for months to treat him to dinner at Spaghetti Warehouse and then to the tour of *Phantom* at our downtown theater. He liked the show, and we sang the soundtrack in ridiculous opera voices on the drive back to my house. Our favorite was "Think of Me."

Think of me, think of me fondly
When we've said goodbye
Remember me, once in a while
Please promise me you'll try
When you find that once again you long
To take your heart back and be free
If you ever find a moment
Spare a thought for me.

I shake my head and pull out my phone. A week before I left for New York Ryan texted me. He was coming back to Ohio for Christmas, was I? It felt so good to tell him no, that I had booked a job in New York for the holidays. Fancy that. See? I am living my dream!

But the *Phantom* images pull me back to him. We only had two years together, but so many memories are saturated with him. I pull out my phone and hover over Ryan's name. Suddenly, my phone rings. Startled, I toss it on top of a pile of T-shirts as if it had turned into a snake. Without touching the phone, I lean over and see the caller is my mom.

"Hi, Mom!"

"Hi, sweetheart! We are all here at Nana's. Missing you. Just wanted to tell you hi!"

I hear a clashing of forks and knives and happy conversation in the background.

"Shhh! Everybody—ready?" says Mom to the voices in the background. Then she counts, "One, two, three ... " and the background voices—all my family—shout in unison.

"Merry Christmas, Lindsay!"

Tears threaten, but I smile. I shout back into the phone, ignoring the startled and unpleasant looks from other customers.

"I wish I were there, guys, but guess what?" I fill them in on the fancy apartment in Hell's Kitchen, Times Square, and the ticket scalper, adding all the right sparkle and shimmer to my big New York Christmas. It comforts me to notice all the good. There is a lot of it.

Nana takes the phone from Mom. "I miss you, Lindsay. But I'm so, so proud of you," she says. I wish I were there to smell her rose-scented clothing. I tell her I will continue to have a great time. I will come back with a suitcase full of stories to share.

When I hang up, Anna twirls me around into a hug. She pulls back from me and squeezes my shoulders. "Let's go."

The theater tickets are legit. Our seats are two rows back from the orchestra pit. *Beauty and the Beast* is a balm to my spirit. Every ounce of the show fills me up. For dinner, we grab giant sub sandwiches at a little Italian deli and indulge in huge slabs of cheesecake. We nap at the apartment until eleven o'clock, then bundle up and shuffle back out into the cold. Anna has to go to midnight mass.

We approach St. Patrick's Cathedral and find a huge line has already formed.

"Will we get in?" she queries. Her eyes are full of panic. She has been calm for most of the day. We have both talked about missing our loved ones, but also found laughter at the show and across the table from one another at the deli. It is strange to think that out of my lifetime of Christmas memories, I'll always have this one with just her, the city, and me.

As we inch toward the giant doors, Anna starts to cry.

"We're going to get in," I comfort her. A little woman with a sharp nose like a finch pokes me from behind.

"You may not. They count y'know. If you were smart, you would have bought tickets."

"Tickets?" Anna wails. "To midnight mass?"

"Of course," says the woman. "It's New York."

I grab Anna's hand and squeeze it tight. When we are only a few feet from the door and can see the light from within, the policeman says he has room for seven more.

We are numbers six and seven.

The choir sings heavenly hymns and we belt them joyously from our pews. The church holds us close despite its massive size. Our footsteps echo through the streets on the way back to the apartment. I curl up and gaze out the window at the twinkling lights. There is no tree this year, no gifts or family around the table. We will wake up to just ourselves and the sweet cinnamon buns we bought in advance. For the first time, the city feels quiet. Anna hums a Christmas carol.

Silent night, holy night
All is calm, all is bright

Silent city, shimmering city, good night.

Christmas day passes in a blur. We sit on the floor of the apartment in front of the TV and feast on cheap Chinese food with a bottle of cheaper champagne. We choose the longest movie we can find at the theater, *King Kong*, and settle in for three hours of distracted bliss with a popcorn and candy dinner.

We are thankful on Christmas night when one of the stagehands texts us with an offer of a ride back to Long Island in his car so we don't have to take the train. The next day, we meet him in Times Square and pile into his small rust bucket. It's a chariot with golden wheels for all the time and energy it saves us.

As we zip down the highway, my phone rings. It's Ryan. The rumble of the wind outside the thin car windows beats in my ears. I decide to answer.

"Hello?"

"Lindsay, it's Ryan."

"I know. Merry Christmas."

"Yeah. Same. Are you happy?"

My heart clutches at his voice and his words. My back tenses as I crouch down low in my seat.

"I am happy, thanks," I say. Anna catches my tone and looks back from the passenger seat with raised eyebrows.

"Well, that's great," he says, but it doesn't sound like it is. He asks about my family.

"They're fine. How's yours?" I don't really want to know. He senses this and changes the subject. He asks if we'll see each other at all in January when I come to Ohio, but I am afraid of the other life that lingers with him. It feels like my options are limited to one of two choices: have a career or have a relationship.

"I don't know if that's a good idea," I say.

"Why not?"

"We'll see. I've got to go." I wonder if I would feel differently if he weren't in the military. I am on the trajectory I am on, and he is on another. They don't intersect, and we can't change, can we?

"Okay, bye." He hangs up and I sit with the phone in my lap, cradling it in my hands. I will not let myself see him during the break.

I can't, because it took every ounce of strength to not end this call with three words. If I see him again, I'll love him again.

I never really stopped.

Choosing Us

THE AFTERNOON SUN WARMS the welcome desk at the Kentucky Museum. I shift in the chair, scanning the large windows for incoming customers as I read snippets from the book on my lap. It is strange to be working at the front today; I am usually upstairs in archives or stocking items on the store shelves. My knee jiggles the book I'm holding. I glance at the clock. It's fifteen after three. Forty-five more minutes. I am ready and I am not.

Fifteen minutes later the door opens and I straighten. Is it him? No, a museum customer. But my book falls to the floor with a thud that shakes the silence. I hand the person a map and point them toward the exhibits. As I reach down to pick up my book, a delivery man enters the building and places a large flower arrangement on the desk. Sunlight glints off the top of the daisies and assorted wildflowers, my favorite.

The man looks at my nametag. "I think these are for you," he says.

"Really?" I ask and peer at the delivery sheet he's holding. I see my name. "Do I need to sign?"

"Please," he says, offering a pen. "It came with a card too," he says and hands me a small white envelope.

"Thanks," I manage as the man leaves. The glass door closes with a quiet whoosh. I sit. Then I stand. I push my book to one side of the flowers and breathe in their light fragrance.

I hold the tiny envelope and tear open the small flap. The card is unsigned but has a message, a question: "Are you as excited as I am?"

I sit down and place the card in the warmth of the sun. I look out the windows.

"Lindsay?" The store manager interrupts my thoughts. "You're due for a five-minute break," she says. I get up and push the chair in. She looks between me and the flowers, but I don't have words for her yet.

"Okay, thanks," I say and head to the bathroom.

For about a year after that tense phone call on a New York highway, Ryan and I rarely spoke to one another. An occasional, terse check-in. Always more words hanging in the air between us than were spoken. Life went on. I was cast in more shows, and I studied abroad in Italy. My sister got married. Ryan completed more training. He moved. I dated someone else. He was dating someone else.

Then sometime around the following Christmas, Ryan's name appeared on my cell phone. I was here, working at the museum, pulling supplies together in one of the back rooms. The phone vibrated in my hand as his name flashed on the screen. My heart beat in my throat. I answered.

"Hey," I said.

"Hey," he said.

"I don't have long—" I started.

"I don't either," he interrupted, "but I wanted to tell you I'm leaving."

"Leaving where?"

"I'm deploying to Iraq."

I remembered crouching by my desk in high school the day the Twin Towers were struck and then, years later, seeing the hole in the ground in New York City. I thought of how I could not find Ryan's face as he marched in his uniform at boot camp graduation.

I pressed the phone closer to my ear.

"Stay safe," I said.

"You know me."

"I do."

"I won't be calling as much now," he said. I nodded like he could see me.

"That makes sense," I said. "With the distance and—"

"No. I'm with—"

"Oh yes. Sure. Her. Great!" I paced the floor of the museum. I knew about her, but I did not want to talk about her. She looked nice. He looked happy with her in the pictures I had seen online. He interrupted my thoughts.

"Linds. I can't talk to you because I owe it to her to try."

I stopped walking. "And?"

"And I still have feelings for you," he said. "That's not right. Calling you when I am with her."

I paced again. I thought he was annoying. And that he was trying to be respectful of her—and me. "Okay," I said. I walked quicker. I had to say it. "I still have feelings for you, too."

"I know."

I felt us hold our breath. I felt us wonder if we could have been. If we could be. We said goodbye and hung up.

After Ryan told me he could not call me, he kept his word. Until two months ago.

The call came in as I was walking outside to join some friends. I was working at a summer theater, rotating performances of *West Side Story* and *High School Musical*. My feet froze on the path when "unknown caller" appeared on the screen. I believed I knew this "unknown caller."

"Hello?" I answered.

"Hi. It's Ryan," he said, his voice farther away than the miles between us.

"How are you?" I asked, and he sighed.

"Some crazy stuff has happened over here. I thought talking to you might make me feel better," he said. I squeezed my eyes shut.

"Are you okay?" I asked.

"I am," he said, but he did not offer anything else.

"Do you want to talk?" I asked, opening my eyes to the night sky. I wondered what the sky looked like over his head.

"I'd rather just listen to you," he said.

"All right," I said. I turned and walked back to my cabin. I don't remember what we talked about, but we talked. We talked that night and for the rest of the summer. On a late evening in August, he told me he was coming back to the United States. He told me he had ended things with the other girl.

"Can I come see you?" he asked.

"I'll be in Kentucky. You want to come all that way?"

"I do."

In the museum bathroom, I splash water on my face and press the cool tips of my fingers against my cheekbones. When I come back into the lobby, the manager raises her eyebrows at me and the flowers again.

"Yes, they are from him," I say and glance at the clock. Ten minutes to four. She follows my gaze and grins at me. She knows. Like most small workplaces, ours doesn't hide many personal secrets.

"Why don't you get out of here?" she asks.

I start to protest, but she says she's got it covered. I gather my book, my bag, and the vase of flowers.

The museum is on campus at Western Kentucky, and the walk to my apartment is not long, just a few blocks. I cannot stop scanning the street, looking for him or his car. What will he look like? Smell like? Be like?

It has been two years since Ryan and I have seen one another in person. The daisies tickle my chin and I think about his message. Am I excited? Yes. Am I terrified? Yes.

When I enter my apartment, I place the flowers on my coffee table and change out of my red polo work shirt just as I hear a knock. As I walk toward the door, my feet float and drag across the carpet. Will I still love him after two years? Do I want to love him? Do I want him to love me?

I open the door to every old answer and new question.

~

Ryan pushes the grocery cart as we walk through the store. We have lemons, onions, and chicken, and he is looking for Gatorade.

"Found it!" I say, holding up flavors until he points to the one he needs. "What do you call this dish?" I ask.

"It's Chicken a la Kerikhin," he says, and he tells me it was created by one of the contractors at his site in Iraq. He is going to re-create it at my apartment for my friends. Gatorade is apparently a key ingredient, because it was the best source of citrus flavor available at their deployed location. I'm touched by the effort he is making to share a piece of his recent life experience with me. A meal is one of the few things he can share.

Ryan's specialty is a classified job; he cannot talk with me about what he does. I remember two years ago when this felt like a hurdle between us. It was hard for me to feel invested in what he was doing. I could not relate to or ask questions about his current work. It felt like we were living in two universes. Maybe that was selfishness? Or maybe it was more than my nineteen-year-old heart could handle at the time.

When we get back to my place, I watch as he moves through the kitchen. He is both familiar and foreign.

"Can you cut these pieces for me?" he asks, and I move next to him, dicing chicken. We work with the Rat Pack playing in the background, and as Sinatra croons we set up the porch with a round table, plates, chairs, and the flowers he sent me.

As my friends pour in with a few offerings of wine, cheese, bread, and dessert, they each give Ryan a hug or a handshake. He fits right in. His eyes are warm, smiling at me over the top of his glass. I've missed being looked at that way. I blush as he holds my gaze. I have missed looking at him.

We trade stories, pour drinks, and laugh long after the stars twinkle above us. My friends help us carry the dishes inside, but we stop them from cleaning up.

"We'll do that, thanks for coming," I tell them and say goodbye. I feel the tug of what we've left unsaid pulling me toward Ryan. He is filling the sink with warm, soapy water. I come to help him, rinsing and drying the plates. He dives in.

"So why?" he asks.

"Why what?" I ask, but I know. Our break up was not agreed upon. I look up at him, his hands scrubbing chicken remnants from the baking dish. His brows are furrowed, and when his eyes meet mine they are tangled with hurt, and yet I see his care for me there too. I am not sure if I can even begin to explain how two-years-ago-me behaved, but I will try.

"Ryan, I am sorry," I stare at the dish in my hand, wiping water from it, making it a clean slate once more. "But I am also not." I feel his body shift away from me. "I was scared. We barely saw each other that first year after high school. We weren't growing together anymore."

"We could have," he says.

"Maybe," I say. "But I wasn't ready to move with you."

"You weren't ready to move for me," he says.

I nod. This is true. When Ryan was stationed in California, he asked me to move to be closer to him. I couldn't see beyond what I was doing then. I did not want to leave college or transfer to a different school. I could not fathom how to make us work when I still had my dreams to pursue. How could I honor this need in myself and also learn how to be there for someone else? I decided against the move.

We finish the dishes and stand apart.

"Let's sit down," I say. We sit on opposite ends of the couch.

From where I am sitting, I can see the black and white checkered box that holds my "Ryan memorabilia." Even after our split, I couldn't bring myself to throw away what felt like proof of what we had. That it was once real. The box moved from my freshman dorm to my sophomore room. Junior year, it had a special spot under the bed in my apartment, and I unpacked it into my closet this past summer when I moved into this place, headed into my senior year.

Underneath the black lid are all of his letters, ticket stubs for events we went to, and our first champagne cork. Before he came today, I combed through each piece like a detective hunting for clues. *What did I choose to lose? I wonder if we can regain any of it back? What if there can be a new chapter between us?*

Over the next few hours we talk in circles, our voices growing hoarse as we try to understand the old choices we made and the new ones we might be willing to make. We cannot repair all the past pain, but there is a willingness to forgive. We talk about what we hope to do. I have dreams to keep dancing, and Ryan wants to continue his military career.

Who will sacrifice? I once thought it would only be me. But what if it was both of us? What if this life takes two? Isn't that what we have been feeling over the past few months as we've talked on the phone? What we have been feeling this weekend? I do not want to journey forward anymore without him beside me.

As we talk we each move toward the middle of the couch, until finally our knees touch and he picks up my hand to kiss it.

"What do you think?" he asks.

I look down at his hand in mine. "I think we are worth another try."

"I think so too."

"I don't know the first thing about military life and how this is going to work."

"I don't either," he says. "But maybe we don't have to know."

I rest my head on his shoulder. I want more answers than we have. But maybe love is like this, two people working for each other's common good without having enough answers. Maybe it does not need to be a duel between us, but instead, a duality. Being together can be both risk and reward, light and darkness, a soft place to land and ground to fight for.

Together and Apart

SIXTEEN MONTHS LATER, I open my eyes from sleep, allowing the first quiet light of an early January morning to greet them. With graduation behind me, I'm living in Colorado, working my first post-college job at a theater in Fort Collins and living in the cast house with two roommates.

I turn over on the pillow and feel my engagement ring rub against the soft sheets. My thumb reaches up to pull it back down on my finger. As of four days ago it is not just an engagement ring. It holds both places: engagement and now marriage.

While I was finishing college, Ryan finished a second nine-month deployment to Iraq. He has been stationed in San Diego for three months, and in that time we've made a lot of flights back and forth between Colorado and California. Weighing my options after college, I chose this job to be close enough to spend time with him. It was important to me then and even more important now to be near him.

I reach across the twin mattress for Ryan, but the place where he was sleeping has grown cold. I shoot up from bed, tossing the covers aside, and I force my legs to walk, not run, down the hall. My body relaxes when I find him. He would not leave without saying goodbye, but still, my heart squeezes in my chest at the thought of the nine months ahead of us. We are married but now we will have to endure another deployment—another nine months apart before our life together really begins.

Ryan is dressed and at the kitchen window, steam from his coffee pooling around his face as he breathes it in. I look at the clock. One hour left. I am ready, and I am not.

Seeing me, he offers, "Coffee?"

I nod and he fills a mug for me. My face becomes blanketed by the hot steam, a small comfort in what feels like an awful morning. I look over to the front door and see his black duffle. He is going back to California today, and in a week or so he'll board a plane and depart for his third tour in Iraq.

I am grateful he is not the only one leaving. Most of my suitcase is packed as well. Tomorrow, I fly from Denver to New Orleans to begin a contract with Carnival Cruise Line as a dancer on board the *Carnival Fantasy*.

The timing is serendipitous; I know this. I have always wanted to try working on a cruise ship, and what better time to be on an adventure than while he is deployed? But even as this goal comes to fruition, it feels like we are on pause.

Ryan and I try to make light of it, laughing at this full-circle moment. We met on a Carnival ship five years ago. Now, we're married and I am dancing for the company. It's like a plan coming together, but no one has taught me how to sift through the hard parts. No one has told me how to live my dream while waiting for us to begin. Maybe no one will. Maybe this is a dance Ryan and I have to figure out for ourselves. We have been long-distance boyfriend and girlfriend; how will a long-distance marriage be different? I don't know yet. But it feels different.

We lift our mugs and sip, then he puts his down and takes mine, placing it on the countertop as well. He pulls me into a bear hug and I breathe him in, my husband. He kisses my head and pulls me back to look at him.

"We're doing this," he says.

"I know," I say.

"I am excited for you," he says.

I want to say I am excited for him too, but I am not. Not really. I am worried this third deployment will break the charm. Maybe our luck will run out and he won't come home this time. I pull his arms back around me, trying to hold onto the joy bubble that has been ours for the past couple of weeks.

Ryan proposed the day before Christmas Eve, while we were in Ohio with a group of our friends and his family. Later that night he broke the rest of his plan to me.

"We should get married before I deploy again."

"No way."

"Why not?" he asked.

I told him I wanted our wedding to be with everyone we loved. "Just like today."

The warmth that had surrounded us upon this future promise to one another felt precious. I wanted the same warmth on our marriage day. While this marriage was between us, we were also joining our two families. I wanted to walk down the aisle toward Ryan with my mom and dad beside me. I hoped to dance and laugh with his mother and father. Raise our glasses to a future of adventures and growing our own family. It felt like a betrayal to begin our lives together without them.

"It is not just one day, though, Lindsay. This is our life," he said. He told me our marriage extended beyond rings on our fingers and a ceremony.

"I know that, Ryan," I said.

"I want you to be my next of kin. I want you to have access to all I have."

On Christmas Eve, as we drove to celebrate with my side of the family, Ryan talked about the impending deployment. For the first time, he mentioned his will. I looked at him from the passenger seat and pulled my coat closer.

"Why are you talking about this?" I asked.

"We need to talk about it," he said.

I told him this felt too big. I did not want to combine the thrill of being engaged with a discussion of his wishes after death. He told me I was promising my life to him. He was promising his life to me, and we would not have a guarantee of how long that would be.

Is this what it means to be a military spouse? Holding the space between being together and apart, living fully and yet knowing my husband's life is promised to the Marine Corps first, not to me?

Over the rest of our holiday in Ohio, we researched what we would need to do to get married in Colorado and had deep conversations with both sets of parents. I think we were searching for wisdom—and their blessing. At my parents' house, I remember sitting on the sofa with Ryan next to me as we shared our final plans.

My dad put his hand on Ryan's shoulder and said he believed we had thought this through, and he and my mom were happy to support us.

"But while a marriage is yours," he said, "a wedding is for everyone who wants to walk this celebration with you. Promise me you will allow us to do that."

Tears ran down my cheeks as I understood what he meant. Ryan and I were going to pave our own way, but our family would continue to be our place of rest. They would continue to offer what they could when they could. To fill us up and let us go.

We made our plan to get married in Colorado, promising we would have a renewal of vows ceremony with all the wedding trimmings with our families in Ohio at a later date. It felt like the best of both worlds, and we leaned into the excitement of having two opportunities to celebrate our union.

A week after our engagement we were in the snow-covered parking lot of the courthouse in Colorado, marriage license in hand. As Ryan and I started to walk inside, the door of a white van parked several spaces up from us slid open. A delivery man stepped out.

"Ryan and Lindsay?" he asked.

"Yes?" Ryan stepped toward him.

"I have these for you," he said and pulled out a bouquet of red roses and white daisies. He handed those to me and then reached into the van to pull out a small container that held a matching boutonniere for Ryan.

"Who are these from?" I asked.

"Let me see," the man said, looking at a clipboard on his seat and running his finger down the sheet to trace the order. "Uh—Brad and Wendy Swoboda," he said.

Ryan's parents. I clutched the bouquet tighter.

"You two getting married today?" he asked. We smiled and said yes.

"Congratulations!" he called as he put the van in gear and drove away.

Inside the courthouse, I pinned the flower to Ryan's chest. We set our coats on the back wooden benches and hugged our two friends from the theater who were there to be our witnesses. Even though the rest of the seats in the courthouse were empty, I felt full. We were doing this. I thought I would still feel upset without our family surrounding us, but as I watched Ryan place himself at the front of the aisle, I was at peace. I placed myself at the other end, facing him. He turned back with a smile, the one with his lips closed. The smile he holds for me. He was waiting for me. I inhaled the sweet gift of the flowers and walked to Ryan.

We were choosing to begin without knowing the ending.

In the small kitchen, Ryan releases his arms from around me and goes to the couch. He picks up his coat and hat and puts them on. I place our mugs in the sink and look out the window.

Big snowflakes are starting to fall, softly covering the ground. I see the lights of another small white van.

“The shuttle is here,” I say. He nods and picks up his black duffle by the door.

Four days does not make a marriage, but it is a start.

We embrace one last time. I try to hold the feeling of his arms around me. I try to breathe in his presence. It will be another nine months before I have it. He kisses me and opens the door. I don’t take in what we say because, at this moment, no words are enough. I return to the window to watch him board the shuttle. It drives away.

I brace my hands on the countertop, and a chill runs through me. Just because we are physically apart doesn’t mean we aren’t together. I twirl the ring around my finger. It will hold his place.

Our Unknown Future

RYAN OPENS THE DOOR to the small apartment in San Diego. I don't know what I expected, but it wasn't this. A small, dark space with one bedroom, cheap cabinets, and mud-brown carpets. A tiny cement balcony overlooks the parking lot. I can smell the black tar from two stories up.

I walk to the bedroom, which is full of cardboard boxes and gear. There is no furniture. Why would there be? Ryan has just returned from Iraq, and I've just flown in from my last days at sea with Carnival. San Diego is our meeting spot before we fly to our belated honeymoon in Australia. When we return, we'll drive up to look for an apartment at his next post in Monterey, California. Then, we'll fly to Ohio to drive my (our) Toyota Yaris across the country.

This apartment is a temporary dump site for our belongings while we work out what is next. We own suitcases, the clothing within them, and tissue-paper-thin hopes and dreams. The last nine months were not bad. But they weren't great for us either.

Between Ryan's being stationed in the middle of a desert and my sailing through the ocean and docking in New Orleans every three, five, and seven days, our communication has been sparse. We talk over the phone maybe once a month, have had a handful of shoddy Skype dates, and create a tiny email trail. I called the family readiness officer for the battalion once, but when she found out I didn't live in San Diego, she told me there weren't any support events I could attend. I was the one who chose to work a dance contract instead of living on base; I was the one who would need to figure out how to endure.

"Where are we going to sleep?" I ask.

"I have some pillows and a blanket. I thought we would want to hunt for the mattress together."

"Okay, that makes sense," I say, and it does. But I am starting to grasp that we are at square one here. We have to build this new story together. Piece by piece. Maybe I should feel freedom in beginning with a blank slate, but instead I feel untethered.

As a kid I played The Game of Life board game. I would fill up my car with an imaginary spouse and children and follow the structured path around the board. I collected cards and did what they said. None of the cards ever guided me to: *Reintegrate with your spouse after nine months apart: collect 1,000 good merit dollars for holding off your selfish grumbling and add 500 bonus points for sleeping on the floor for your first night back together.*

I look in the kitchen and open a few cupboards. They're empty.

"We don't have anything to cook with," I say.

Ryan sighs. "I know."

I side-eye him. I don't want to sound disappointed or ungrateful, but I honestly don't know how I feel. We've endured this time apart and yet being back together is not as simple as I had hoped. There is unseen work to be done, and much of those needed repairs will take time. We cannot rush it even though I want to.

Ryan comes behind me and opens the fridge. He pulls out two beers, walks across the small room, and slides open the glass door to the tiny balcony. I follow him out.

"These are nice," I say as I sit down in a green camping chair. I smile. I need to help turn the tide. "Our first furniture!"

I raise my beer, and he clinks his bottle with mine. I reach for his hand and trace my thumb across the top of his. This moment does not feel real. After waiting so long to hold his hand, now I have it. That matters more than dim apartments, bleak parking lots, and brown boxes. I take a sip, and the smell of hot black tar mixes with the orange wheat beer, fresh and acrid, tangy and sharp.

That night, as we lie on the mud brown carpet with our one blanket and two pillows, I tell him about how I taped pictures of him up on the ceiling of my bunk on board the cruise ship.

"Each night as I went to sleep I would stare at you, because sometimes when I did, I would dream about you," I say. I tell him it would brighten the whole next day, feeling like I had seen him and spent time with him. I cringe a little when I hear myself admit this out loud. Does it sound too much like a Nicholas Sparks novel? I look over at him, and Ryan is staring up at the white ceiling above us, his hands folded onto his chest.

"Is that weird?" I wish I could suck the words back into my body, keeping them safe.

"No, not weird," he says, but dreams can only go so far.

I nestle my head deeper into my pillow and close my eyes. It is time to build a life beyond the board game. We have to work, not wish, to form a reality that looks like us, in our marriage, knowing each other in real life. I realize our days ahead will not be full of sprinting to a Skype call or pressing our cheeks up to phones, listening from miles away. I have grown more accustomed to that part of our story. I know those rules. But it is time to learn new ones. What is every day like now? What are we like in person together instead of the dream of being together?

This unknown future is our unknown future.

~

I position myself in a square of sunlight gleaming in from a window of our apartment in Monterey. We've been here for three months, and I don't miss the San Diego transitory launch pad. My phone sits to the right of me, a notebook on my left. The pen is lying on the carpet. The lines I scrawled on the paper before throwing the pen to the floor are a betrayal, an exit strategy, or a turning point in my career. I don't know which. I need to clear my head, so I pull on a jacket and walk out, locking the door behind me.

My feet fly down the terra-cotta tile steps and through the gate of the white picket fence. I can smell the salt on the air from the Pacific, and I watch as a gull careens down the streets of Spaghetti Hill in front of me. An airy, light-filled gem, our apartment is minutes away from the main gate of the Presidio of Monterey, where Ryan is a student at the Defense Language Institute. He cruises up to his classes on his new yellow sports motorcycle, and the blue Yaris waits for me in our driveway.

Waits for me to go where, exactly? I think as I shove my hands in my jacket pockets and curve down our block. We will have to talk when he gets home, even though I don't know what to say. We haven't had many disagreements yet. Our delayed honeymoon had been everything we had imagined it would be. But I feel our honeymoon days are fading, and this new life with Ryan in California is not quite as I imagined.

What we had last night was more than a "disagreement." It was an argument. A shouting match. A verbal shoot-out. I ended up sobbing in the garage, sitting on top of the dryer. I wince as I remember the advice we were given by friends and family at the start of our marriage. Advice like, "Never go to bed angry."

But when I came up from the laundry room, we didn't talk. I washed my face and crawled under the covers of our bed. When I woke up, Ryan was gone.

It's not like I haven't been trying. I've hung photos on our walls. I go to group exercise classes and spouse coffees. I have made lists of different dance studios in town, including places that offer Zumba fitness programs and ballroom dancing. I even made a Thanksgiving feast for two this year. Ryan and I go to the weekly farmer's market, play Frisbee in the park on Saturday mornings, and are looking at joining a wine club.

But I haven't danced since August.

One night on board the *Fantasy*, in the middle of a performance, the music stopped. The house lights came on and an announcement brought the show to a halt. As we dancers skittered off into the wings and threw on our sweatpants, we heard the story. A teenage boy and his dad had argued. After the argument, the boy stepped out of his shoes. And then he jumped overboard.

I remember walking out onto the crew deck that night. The ship was halted in the water. I watched as the search and rescue boats lowered. Giant beams of light swept across the water. For miles and miles, the water was as still as glass. The ship was one speck in the middle of a dark nothing. I imagined the boy must have been like a speck within a speck, like the Whos in Who-ville on a dandelion puff, shouting for Horton to hear with his giant elephant ears.

The silence was louder than sound, like when a character in an action movie gets cotton-ball hearing after an explosion. I kept waiting to hear shouts of victory, of rescue, but they never came. I stood with the stars above me, thinking of Ryan and wishing I could call him and share this sorrow instead of carrying it alone.

What had made the boy jump into the darkness? Why had he left his family with only their memories and his last pair of shoes? Eventually, the engines shuddered back on, the search boats came back, the great ship heaved, and we sailed away without the boy.

My walk carries me away from the apartment, down to the water, and I scuff my feet on the rough wooden planks of the dock. I want to live with my feet grounded in what matters most to me. I just didn't know deciding to be with someone also meant being called into my own battle. A battle to fight for a new direction and perspective, to examine what I value.

The ocean laps at the dock, and I think back to our fight.

"Why can't you be happy?" Ryan had asked, running his hands through his hair. "We did all of this. We're here, and—"

"I don't know!" I'd screamed. "I don't know!"

And I still don't know.

Maybe part of this first year together is to figure it out. I am mad that we're finally here, we have taken this leap of faith, and yet I still feel like I'm missing something.

Another gull swoops overhead. I hear the soft swoosh of its wings as it banks down to a patch of sand. I lean onto the dock railing. Since Ryan and I moved to Monterey, I have been mourning the career I left behind. While I was ready to stop performing, I was not ready for the blow that decision dealt to my identity. Who am I without the constant flux of audition, rehearsal, and performance?

I start walking back up the hill to the apartment, thinking about the phone call I placed earlier today. My friend at the theater in Colorado had been sure that there would be a slot for me there if it was what I wanted. I sat on our couch and scrawled down the dates for rehearsals and performances. I would be gone for eight or ten weeks in total. I rationalize: Ryan and I had talked about the possibility of my taking other performance contracts.

I turn my key in the lock of our front door, and I sigh. I could run to Colorado, go back to performing. Is that what the blue Toyota is waiting for? Would I drive it to Colorado? But then Ryan wouldn't have a car here. And I would not be taking the job for the right reasons. I hear the familiar whine of Ryan's motorcycle as it turns down the alley. There is a clack as he downshifts and coasts into view.

One person is not responsible for my happiness either.

Ryan and I talk that night. Our voices and bodies lean toward one another as we seek understanding. I pace the kitchen while I tell Ryan about the potential job in Colorado. He leans on the countertop and tells me he never expected me to give it all up.

"Do you want to go?" he asks.

"Of course I do. But we just got to the place where we are together," I say. I tell him I know I don't want to go back to where I was, chasing the familiar because marriage feels hard right now. But I am also scared I cannot see what is in front of me.

"It does not feel enough right now to just be—"

"My spouse?" he asks.

"Your dependent," I say.

His jaw clenches and unclenches. "You are putting hurdles up where you do not have to. No one is asking you to do that."

I nod. The word "dependent" is rife with my assumptions as I start to understand military life. Does being a dependent mean playing dress up and being a happy homemaker? Does it mean I'm sitting around leeching the benefits of my spouse's career? Am I dependent on some outward thing like a relationship, a career, or a title to make me happy? With all the time we have lost to distance and deployments, can I depend on this time together to heal us so we can move forward?

Or does dependent mean nothing at all? It's just a word. A word doesn't determine my path—ours. We choose our path.

We talk until we have a plan. I will stay in California with Ryan. We will look for opportunities here, beginning with figuring out what I might want to work on next.

Over the weekend, we sit down and talk through different career options for me, looking out over the next few years. We talk about furthering my education, new certifications, and updating my resume. I tell him I believe I am ready to take on more of an educational role with theater and dance.

"But I also miss dancing just for me," I tell him.

"I don't know if I can help you with that," he says, and I grin.

I do have an idea for this part of the plan.

Learning New Steps

THE BALLROOM FLOORS AND MIRRORS are polished, allowing the twinkle lights along the top of the mirror to shimmer off numerous surfaces. I put on my LaDuca character shoes, their soft leather hugging my feet, worn in all the right places from hours of rehearsal. I do not own ballroom shoes, so these will have to do. Ryan reaches for my hand and pulls me from my chair. The instructor calls all couples to the floor. We take places near the back of the room and start to learn the beginning footwork for salsa.

In another thirty minutes, the class is spread out so we can rotate partners. Ryan raises his eyebrows. I did not anticipate this either, but I can't stop smiling as I watch Ryan rotate around the circle with confidence. He will always be my favorite partner, but the white-haired gentleman with black-framed Coke-bottle eyeglasses has great leading skills. He's a good five inches shorter than me, though, so turning under his arm challenges my knees.

As I change back into my tennis shoes, the instructor approaches.

"Have you danced before?" he asks.

"I have," I say with a smile. I glance over at Ryan, and he gives me a slight nod of encouragement. "I have my resume with me. I don't have experience with traditional ballroom dancing, but I have done a fair amount of other training," I say.

"Please, I would love to see it," he says. I offer it to him and he takes it to the front desk to review it. I pull on my light jacket and

Ryan lends me his arm. We stop by the desk on our way out.

"I see here that you taught a series of elementary school classes?" the instructor asks.

"Yes, for a program my college assisted in with local schools."

"Did you enjoy working with children?"

"I did. I do!" I say and explain that many members of my family are teachers. "They might not be dance instructors, but I believe I was raised to have the natural heart of a teacher."

"We may have a position for you if you are interested."

"I would love to hear more!" I take a deep breath as I shake his hand and make plans to meet the studio owner on Monday.

When we get into the car, we are quiet as we begin the drive, but a loud silence fills the vehicle.

"I know you want to say, 'I told you so,' " I tell Ryan.

He pats my leg as he turns the car up onto our street.

~

While it feels like the weather in coastal California stays the same each day, subtle shifts in the growing season mark time for spring. Ryan downshifts the motorcycle and we swing into the parking lot of Heller Estate, a local organic winery. We are ready to participate in a ladybug release party and enjoy delicious food with a sunset view of the valley.

I climb from my crouched position off the back of the motorcycle and my leg muscles give a silent yelp. They are sore from the busy but satisfying teaching schedule I now have each week. I split my time between teaching for Dancing Classrooms, a program where I teach fifth-grade students basic ballroom dancing to encourage culture and community, and subbing at a performing arts academy in Carmel-by-the-Sea. I have my fingers crossed for a full-time position in their summer camps and for next school year.

Ryan and I walk to the check-in table and we each receive our drink tickets, a cup of ladybugs, and a spray bottle of sugar water.

We are instructed to spritz the leaves of the vines before gently placing the ladybugs down upon them. Up and down the rows we go, releasing the ladybugs out to fight the bad bugs of the vineyard.

I watch insects dodge our spray and think about how I wanted to retreat when our marriage felt more bad than good. I did not know how to stay and fight this past autumn. I have a twinge of regret, but it doesn't last long. Ryan and I are beginning to accept that we both carried hurt into this marriage. We have to sift through our pasts to determine the healing of our present and future.

I wanted to think this duty station was our finish line. After the past three years of our long-distance relationship intertwined with deployments, we made it to marriage. I wanted to heave a sigh of relief that we were "done," but being together in person is not all it takes to build a marriage.

A ladybug crawls from the cup onto my pointer finger. The feet are like a tiny feather tickle to my skin. I guide the red-winged friend to a leaf and remember ladybugs can be signs of good fortune. Ryan walks ahead of me with his cup and spray bottle. His strong shoulders straighten with confidence and skill. He is meticulous and kind as he lays the little beetles across the foliage.

I look at him and think, *What if this time is our starting line? What if we continue to set the pace together?*

He motions for me to follow him but I say, "You take that side, I'll take this one."

"Sure," he says.

"I'll meet you at the end of the row."

Ryan and I continue our mission as the sun tilts toward evening. Another ladybug drifts down my palm to the grapevines, and I turn over the definition of "depend" in my mind. The infinitive "to depend" has more than one meaning. One is "to be controlled or determined by." Another is "to rely on."

We reach the end of our row and walk up to the party to enjoy appetizers and collect the two new wine bottles from our membership. I wrap them carefully in the towels we brought with us and place them securely in our backpack. Ryan climbs onto the motorcycle and I follow, leaning in close and locking my arms around his waist. I hold tight as we turn on the highway toward home. I can rely on our continuing to grow, to champion each other, to choose love during and after anger.

The road widens before us, and I rest my cheek between his shoulder blades. I did not give up on my dreams and ambitions to be here with Ryan. But I did trade them in for a new route. The night sky blazes with stars above us, and I soak in the warmth of Ryan's back, the afterglow in my heart from a good experience together, and the peace of knowing I am right where I need to be.

Starfish

I FIND COMFORT IN THE STEADY THRUM of ocean waves. I love salty air and how the sand shifts beneath my feet. When I am in wet sand, it sucks me down and begs me to stay just one more moment until the water laps at my ankles. When it is white and sun-baked, it emits an audible squeak with each of my steps. I imagine the sand stretching and yawning as I walk across it, awakening it from a sunny slumber.

Mom and I walk the beach with our arms linked. Her warmth next to me makes me smile. It is hard to believe it has already been several months since she and Dad linked arms with me and escorted me down the aisle for our long-awaited wedding celebration in Ohio. I squeeze Mom's arm in comfort as we walk. We prefer the wet sand and we tease the waves, walking just close enough to pad gently in the surf. Mom's navy-and-white-striped zip-up sweater is drooping in the front. Its pockets, heavy with sea treasures, stretch down to slap her thighs. I steady her as her fingertips reach down to the beach. She is a master treasure hunter. She always has been.

I remember Mom teaching me how to seek shells when I was a little girl, around age eight, visiting Myrtle Beach for the first time. I became a hawk, scanning the sand to surprise my mother with the finest ocean art. She taught me to look for texture in gray bits of coral and to dig out delicate shell fans. Mom's favorite creature has always been the sand dollar. We hunted for the brittle white mystery, ever hopeful we would find a complete one. I have always favored the starfish, with its spiny, reaching legs. I love knowing a starfish can regenerate a new limb if it loses one. It seems like the epitome of self-sufficiency.

There is also magic in having stars to wish upon at the top of the heavens and down in the deep-blue sea.

In our family, it was Mom who made sure vacations happened. When she was planning an adventure, I would watch the steady excitement she maintained, from researching a place to slinging bags into the back of a car, taking joy in the experience. Once a year our family escaped from the landlocked, cow-covered, cornfield abundance that is Ohio to explore East Coast beaches within driving distance in our big white Chevy van. Our summer days were spent on the beach, swimming and collecting shells. For dinner we would have crab cakes and lobster, boiled shrimp tumbling across plastic table cloths with potatoes and ears of corn. We spent spring and summer breaks with wet hair, wide-brimmed hats, red checkered bibs, and butter sauce dripping down our chins.

My sister and I would sort through and choose our favorites out of all the pieces of shell and rock we gathered on our forays. They came home with us to live in glass jars on our bookshelves, a reminder of sandy toes and salty hair.

As Mom bends to grab another shell, I realize I am in the role of tour guide and trip planner this time. She and my dad are here to visit during their spring break from school, and I'm introducing them to Monterey. This morning, it's just Mom and me on the beach.

The Pacific Ocean is frigid, but we still dare to dip our toes into the choppy surf. The sand here is not pristine white but brown-black, scattered with lumps of kelp. Black rock formations cut into the coastline, and harbor tidal pools are full of treasures. With our arms still linked, Mom and I climb our way back up to the road. She empties her heavy pockets into a small canvas bag and ties it securely to the black metal basket on her bike. We pedal up the hill until we see my apartment, its buttercup hue distinguishing it from the others. We have to stop and jump off as we inch closer, walking and pushing the bikes up the last steep parts of the hill.

On the porch, we lay out the bounty: bits of shell, rock, and oyster halves. The wind gently rustles our hair and carries with it whiffs of chowder from the restaurants on the wharf. We can hear the excited bark of sea lions as I carry out a bucket of water to rinse the treasures.

"I took a few photos of starfish for you," Mom says, rubbing a clam shell between her thumb and forefinger, pausing to scrape a layer of sand from the shell with her fingernail.

Mom's necklace sways away from her neck, a silver charm in the shape of a sand dollar. She loves the sea so much she always takes it with her.

Ryan and Dad understand our need for girl time, but the next day's hike will be for everyone. Ryan and I have lovingly dubbed Monterey our Foggy London Town. The Pacific is unpredictable. Storms roll up last minute, or the blazing sun will break through and surprise our skin. At times we've arrived home with both wind and sunburns. We've learned to dress in layers like the locals we feel we have become. Ryan is immersed in his language program for the Marine Corps while I zip up and down the highway to teach dance and fitness classes. In these early years of marriage, our love rises and falls like the tide.

On our family hike, the Point Lobos State Natural Reserve offers up the coastline on a splendid platter. We walk along a trail that weaves in and out of pine needles so soft it feels like walking on carpet. The cypress trees gnarl their way up to the sky, and around each bend we gaze off into the ocean. Mom spots a bench looking over the sweetest vignette of trees, rock, and frothy surf.

"When I die—or when I'm about to—wrap me up and stick me here!" she laughs. I see the glimmer of an older woman in her eye.

Though Ryan and I are writing our own family rules now, our parents remain our anchors in deep waters.

Three months ago Ryan and I had pedaled across the bike path from our home to the strip mall in Seaside, chatting about the move we knew was coming, though we weren't sure where. Our time in Monterey was ending. Something new was on the horizon.

"This place," I said backward over my shoulder to Ryan, "it's been good." I looked out across the endless stretch of the ocean and remembered how nervous I had been about coming to California.

"How do you mean?" Ryan said.

"I mean, it's all worked out. It's been really amazing. I know we can make it work anywhere." I stood a little on my pedals, feeling the strength of my legs.

"Anywhere, like Korea?" Ryan asked, pedaling up beside me. I looked at him.

"Because that's where we are going," he finished, pedaling away again, perhaps faster than necessary to avoid what could be my fury or my excitement or a collision of both.

I remember my head snapping forward and almost ramming into an overgrown bush on the path.

Korea.

I didn't get the details until we pedaled back home. Ryan told me the orders were stamped. Seoul would be our new home. We would head out in five months.

I hesitate by Mom's put-me-out-to-die bench and watch her link arms between Ryan and Dad. I sink down on the old gray wood for a moment to collect myself. It is hard to think of moving so far away from my family, even though Ryan is my family now. It makes me think of a giant scale, balancing all we will gain and all we will miss. I cannot tell yet which way it tips.

The ocean roars, tossing up an unexpected wave. I watch it crash onto the rocks below.

"Lindsay?" Dad has come back for me. He extends a hand and pulls me up from the seat.

"Are you okay?" His eyes hold mine with an intense gaze.

"Sure. Yeah. I don't know," I fumble, "Will you and Mom be ... okay?" I ask. My underlying worries surge. What if something happens to them while we are halfway around the world? There would be no time for goodbyes.

He lifts his eyebrows and replies, "We are fine. Mom and I support you both. You know we will visit."

We cross onto the pine needles again and they cushion my steps. Dad takes my cold hand and holds it, placing it with his own in his jacket pocket.

The warmth of his hand draws me back to a memory of a conversation we had the previous fall when I went back to Ohio for a visit without Ryan. Dad had taken me out for dinner and a basketball game. At the pub, we talked through the big changes that were happening in my life and about how I was navigating places he had never been. I remember thinking how strange it was to be sitting with him as an adult, but also how he would always be my dad. There are miles of bookshelves filled with books about how to raise babies, toddlers, and teens. But not enough on the delicate dance of how to continue parenting alongside older children. I wondered if the act of parenting slowly shifts to mentoring and friendship.

As the pub filled with people toasting and laughing, he squeezed my hand. "You know I've loved your mom and you girls and our life together," he said. "If anything happens to me, I've lived well. You should know that."

I have replayed his comment as the move to Korea creeps closer. I don't like my parents talking about being anywhere but here on this Earth, ever-present, with me. And yet we don't know how much time we have. My conversation with Dad has stretched over me, comforting me, securing me with permission to leave. And permission to live. Live my own story with Ryan.

As the water crashes behind us, I decide that no matter what happens in the years ahead, I will have these memories. I have to stop missing opportunities to build new ones because of worry.

Ahead on the path, we find Mom and Ryan hovering over tidal pools. There are brilliant purple urchins and scampering crabs. As I cross the wet rocks, I find one basin with starfish stuck to all sides of the rock. They hold fast and let the water wash over them.

We return home wind-whipped and tired, but it is my birthday, which means overdosing on sugar with an assortment of cupcakes. Mom and Dad always let my sister and me choose our birthday cakes. I fluctuated between two flavors: either Oreos or Reese's Peanut Butter Cups. Tonight, I have cupcakes in both flavors. Ryan places candles shaped like the numbers two and five, one on each of two separate cupcakes. He pushes them together to mark my quarter-century milestone.

It is strange and nice to have the three of them sing me "Happy Birthday," the family who raised me up and my new family with Ryan. I blow out the candles with a wish that is soon forgotten. Perhaps it was simply to move well in the upcoming months.

Mom walks to her purse and pulls out a purple envelope. When she hands it to me, I can feel there is more than a card inside. She has not licked the seal; the flap is merely tucked. I pull out the card, a ballerina in a dazzling purple leotard with a pink feathery tutu. I grin because it is a card meant for a little girl in love with ballet. I am a twenty-five-year-old still immersed in and devoted to dance, even though I have hung up the painful pointe shoes for the instructor spot at the front of the class.

Inside the card, Mom has written, "You will always be our star."

Taped underneath is a small plastic pouch. I peel it off and unzip the bag. Tipping it gently, I gasp as a brilliant silver starfish charm on a delicate chain falls into my palm.

~

The spring blossoms fade with the first heat of summer, and Mom comes back to California. This time she's here to help me with final preparations and to see me off on my long journey.

Ryan has already gone to Korea to make his report date. I stayed behind to finish my contract with the performing arts academy in Carmel-by-the-Sea. Last night was the final show. There were bows, goodbyes, tears, and my students with their strong, tiny arms wrapped around me in farewell hugs. I stood in the wings and could see my mom out in the third row of green velvet seats. Her face, lost in the music and smiling with the movement, looked the same as it did when I was the one performing on the stage.

Our buttercup apartment is empty now. I watched the movers box up our household possessions and place them in giant wooden crates. I initialed all the lines on the forms as they nailed the crates shut. Mom and I rented a car so we could drop the tiny Toyota off for shipping. I signed more forms and watched as our little blue car was driven into a metal cargo container.

Now Mom and I are standing together in the Monterey airport, checking my bags. The attendant hoists two huge suitcases onto the conveyor belt. These are free of charge, he assures me, after seeing my military orders. I'm not heading out for a Korean holiday. This is the new mission.

The airport is stark. Gray and white and blank compared to our faces, which struggle to hold strong. Mom links her arm with mine, and we walk to the security checkpoint. She stands as I make my way through the scanners. I feel her behind me. Her hug has left the smell of her perfume on my shirt. I press my chin down into the scent and breathe it in.

Her presence has helped me feel safe to grieve this departure. I have cried over leaving our beach town. Over leaving a job I love, the roads I know, the restaurants we frequented, the bike path I pedaled. I already miss our first apartment, the barking sea lions below and our voices inside, laughing together, fighting for each other. It is not just Monterey I will miss. It is the English language and American soil and the comfort that I could catch a plane and be on my parents' doorstep in four short hours.

The airport is small enough that when I get through security, I can still see Mom. A glass wall partitions us, and I shift my backpack onto my shoulders and meet her at the glass. I want to crash into her like an ocean wave, cling to her, and pull her with me. Instead, I place my hand on the glass. She matches her palm with mine.

"I love you," I mouth, crinkling my eyes into the most reassuring smile I can muster.

"I love you too. See you soon," she says back. Her smile matches mine. As she turns, I see the glint of her sand dollar necklace on her collarbone.

I turn and grab my starfish, clutching the dainty chain and tracing my thumb and forefinger over the ragged edges of the silver creature. Leaving here, leaving her, is like having a limb ripped off. But I will regrow—I can hold on and let go.

Ending or Beginning?

THE RESTAURANT IS NO BIGGER THAN some walk-in closets in America. Ryan and I sit at one of the four tables, awaiting my first meal in Seoul.

"I think you'll like this," he says as he orders. I drink from a bottle of cold water and rub my temples. Was it just yesterday I flew here? Jet lag hangs over me like a cloud. I made it until nine o'clock last night before I crashed in our bed. Well, not exactly our bed. It and the rest of the dented and dinged-up furniture in our two-bedroom, seventh-floor apartment is rented from the US Army. The concrete gray-flecked tile in our hallway reminds me of hours spent in old school classrooms.

The stand-up air conditioner in the restaurant washes over us with welcome surges of air, cooling the sweat rolling down my back. We have a similar unit at our apartment. It plays a song every time I turn it off or on. Ryan told me this is normal, that most Korean appliances sing sweet little jingles. A waiter comes to our table and places a giant black hot pot in front of us along with two bowls and chopsticks. Then he begins ferrying side dishes from the kitchen. Eventually, there are at least ten little white bowls around us, a lot for one table. It feels like a feast for the senses: colors and textures, smells, and what I anticipate will be new tastes, too.

"This is boo-doo-chee-gay," Ryan says, pronouncing carefully as he ladles some of the soup into my bowl. My brain sees it phonetically, as the ramen, cabbage, red broth, and hot dogs slosh down in front of me.

"Hot dogs?" I ask and laugh. "I came all the way from America to eat a hot dog?"

I raise my eyebrows at Ryan, and he explains that this soup is one of his favorites now. *Budae jjigae* was created during the Korean War in the 1950s and roughly means "army stew."

"Food was scarce in Korea at that time," Ryan explains. "So, meals were supplemented by leftovers and scraps of processed food from the American soldiers." Hence the hot dogs.

I watch Ryan as he dives in, and a little lightbulb clicks on in my mind. He is now part of the mission here. We are part of the mission of this place—and its history. I have been so focused on finishing my time in California, the move, and how I was going to handle the change that I never paused to consider the military history our family is joining.

I take my first bite of army stew, cough, and laugh at myself. When I look up, Ryan is grinning.

"Am I bright red?"

"Not yet. The heat and spice are something to get used to."

"There's a lot to get used to," I say as ramen noodles slip past my chopsticks.

"I know. I'm glad you made it, though. I've really missed you."

I look up and study his face. "You've done a lot while I was wrapping things up in Cali," I say. His eyes meet mine. I know I don't have to say it, but I do anyway. "Thanks for letting me have that time."

"It's something we chose together, but eighty-nine days was really pushing it," he says, and I choke on hot broth.

"Was it really that close?"

"It was."

Ryan's orders had been for the summer. When they were moved up to the spring, we agreed I would stay to finish my contract at the dance studio and see my students through the end of their season. To travel on Ryan's orders, we learned, I had a ninety-day window between his departure and my arrival in Korea.

A few months before our move, I attended an on-base class for military spouses to learn more about military life. As we were leaving class, I was talking to my friends about the process of our move to Korea and my plans to stay behind for a few months. In the hallway, the woman who led the class stopped me and asked if she could talk to me for a moment. I agreed.

"I overheard your plan," she said. "That's a long time for your husband to wait for you. You should consider moving together."

My heart started to pound. I felt like a bad student being called out for making a poor choice.

"We weren't supposed to move until this summer," I tried to explain, as if I needed to. "I've signed a contract with the studio to finish the year with my dancers. There are competitions and the recital still ahead of us."

She waved my comments away, as though clearing out a bad smell.

"You can get out of that. It's your duty to be with your spouse."

I don't remember what I said after that, but I wished I had a better comeback than the politeness I forced through my teeth. I remember coming home to Ryan and unleashing my fury about the audacity of this stranger to weigh in on our situation.

After I released everything, I paused and said more gently, "I really want to stay for my dance kids. But do you want me to leave with you?"

"We've talked about this, Lindsay," he said. "I am fine with your staying. It's not great, and I wish you were coming with me, but it's what we have decided is best."

I placed my hands on his face and drew his lips to mine. The Marine Corps holds a lot of our life in its hands, but we do still have some choices. Sometimes those choices will not make sense to anyone else, but we get to make them together.

Now, here I am on day ninety, sitting across from him in Seoul.

"I think our choice was a good one," he says, "don't you?"

"Oh yes," I say, and fill him in on the time we've been apart. I show him recital pictures and share the kind words from the dancers and their parents. I talk about how great my mom was, helping with the pack out and our final beach walks and meals together.

I leave out the part about how I feel a blossoming ache from leaving California. The ache is tangled with the love I have for him. It is our next right thing to be here in Korea. But I already miss the life I was living. I try to shake off the feeling by reaching for the different side dishes spread across the table. Having been here for months without me, Ryan is excited to show me what he has learned. He tells me what each dish is and fills me in on his co-workers and what his past few months have been like.

On the way home he points out little places in the neighborhood he has already discovered. We hold hands, and I squeeze his palm into mine, trying to anchor myself here with him.

"Tomorrow I'll get your ID updated," he says. "Over the weekend we'll get a phone for you, and we can get your subway card."

"That sounds good, babe," I say, and I mean it. He knows I need options. He knows I need a sense of freedom. By enabling me to have all the suitable materials, he is helping me to grow wings again. Of course he is.

As we brush our teeth before bed in our tiny bathroom, I catch his eye in the mirror.

"Thanks for tonight," I say.

He grins and says, "There are more adventures ahead."

"Sure," I say after spitting into the basin. Staring at the green and white blob of toothpaste, I ask, "What do I do next?"

"What do you mean?" He asks.

"What do I do here?" I ask, raising my eyes to meet him in the mirror again.

He looks at me for a moment and doesn't answer. I imagine he is thinking, *of course she is asking this. Of course she is.*

Instead, he says, "You just got here. Give it more than a day."

I click off the bathroom light and follow him to our bedroom. He turns to me and adds, "Maybe the best thing you can be here is yourself."

Jet lag wakes me again at four in the morning, and I creep out of our bedroom and to the kitchen. After pouring myself some water I stand looking out our living room window. Seoul spreads out before me. I sip from the glass of cool water and turn over Ryan's last comment in my mind.

The best thing you can be here is yourself.

That sounds great, in theory. But those words also make me flinch, uncertain who that self will be. In California, I had to find out what it was like to not be a performer anymore. I had to discover what it meant to be freshly married. I also discovered I loved teaching. I am capable of growth. The rewards of those discoveries feel like a return on an investment, but I don't know whether to save this windfall or spend it. I also realize I have no idea where I could spend it—yet.

The city lights begin to wink on inside some of the nearby apartment buildings as the day dawns. Green ginkgo leaves wave at me from their trees, illuminated by the streetlamps below. I take a big gulp of water. What would it mean to not be a teacher here? What would it mean to not work at all? A taxicab beeps, a man walks down the street swinging an umbrella, and I lean my forehead against the cool glass of the window.

Seoul, are you a place that will swallow me whole, or will you surprise me?

Are you an ending or a beginning?

Brief Sabbatical

IT IS A COOL AUTUMN EVENING when I duck underground to catch a subway home. I notice the last orange glow of the sun winking at me. I like how autumn starts to bend the sun beneath the earth earlier and earlier each day, but I don't love the encroaching darkness. The shorter daylight hours remind me that the seasons are changing, whether I want them to or not. I'm sure it will be dark when I pop out to walk the last few blocks home.

I tug at the scarf around my neck. I shift my weight in the middle of the train. I feel hot and itchy with dance-class wet sweat still stuck in my leotard and tights, covered by jeans and a sweater. But it's worth this temporary discomfort to have a piece of my soul restored here in the city.

I shift again as a tall man squeezes past me. I wonder if I stink. I touch my temple and find the salty glaze of sweat mixed with hairspray. Sometimes I still forget to pack a change of clothing as I gauge my journey between the studio and home. Today I got to the studio barely on time after messing up the train transition once and running down the street until I spotted the right building. I rode up the elevator, punched my class card at the front desk, shimmied off my cover-up clothing, and hopped into my jazz shoes and into the studio just as the teacher cranked the tunes.

I found this studio because during the summer I had auditioned for a production with the Seoul Players, a theater group made of expats from all over the city. The open call was held at a bar. Bars are never great in the daylight hours. As I walked into the black hole, my feet stuck to the countless spilled drinks on the floor. A thick haze of old cigarette smoke clung in the air. I considered turning around and walking back out the door. Maybe this wasn't something I wanted to do. But the other auditioners smiled at me and radiated a welcoming warmth. I felt like they could be my kind of people. I received a paper with lines to read, and then I stayed for a dance call.

I felt ridiculous, as no one else was dressed in dance clothing, but the director of the dance piece smiled at me. I remembered Ryan's words, and I chose to stay. I stepped onto the small stage in my tights and leotard and gave the dance audition my full attention and energy. When it was over, the other casting directors told me they would "let me know." I nodded and packed my things.

As I was leaving, the dance director chased me out the door, her blonde ponytail waving behind her. "Hi! I'm Veronica. Can you grab a coffee?" she asked.

"Sure," I said.

While we walked through the neighborhood, she told me I was her first choice for her show. Then we sat down in a small café and spent two hours swapping life stories and planning the next time we would hang out. She invited me to a dance studio she had found, and we agreed on a class to take together.

When Ryan came home from work that night I ran to the door. I told him I had booked a gig, and I had also made what felt like my first solid friend.

"That's amazing, Linds," he said. "I can't wait to meet her."

"You will! You know what's weird?" I asked as he leaned down to untie his boots.

"What's that?"

"The coffee we shared—it was like dating all over again," I said.

"Oh really?" He asked with a smirk.

"Yeah. I think once you are married, then it is friends you date."

"I can see that," he nodded. "So, did she make it to dinner date level?"

"I think that's coming. But next up is a dance class," I answered.

"Well, that is your version of a second date," he said with a wink.

Veronica became the steady friend I hoped for. We attended many dance classes together, won first place in a Seoul Players event called the Ten-Minute Play Festival, and achieved "dinner level" friend date status.

The "jazz-funk" dance class I'm traveling back from has all the elements I love, from a slow-building warm-up to across-the-floor patterns, finishing with an elaborate routine. Veronica and I often attend this class together, and although she couldn't make it this evening, I decided to brave it alone. I don't speak Korean, but in the dance class it is not necessary. Music and movement are a universal language, a place where my body can speak the words I do not have.

For the final routine, the class broke off into small groups, and my body and mind drifted into the familiar feeling of being both fully present and lost in the moment. Many of my days feel like this as well. I am both lost and found. I am untethered by work requirements and a schedule. Endless days stretch out on my calendar, and I can fill them however I like. It sounds nice when I explain it to friends and family back in America: My time is my own. But it also lacks purpose.

I sent my resume to be an independent contractor at Yongsan Garrison months ago. The response so far has been silence. I have learned the Status of Forces Agreement between the United States and South Korea dictates that the only place American military spouses can work for pay is on the base. I am not the only one aggrieved by this situation.

In my first month in country, I attended a group meeting with the Happy Hookers of Seoul, which ended up being a bunch of spouses with crochet needles and a supply of raucous laughter sipping lattes at the Starbucks on post. I gave up on the crochet part after too many attempts at a snarled hat project, but the honest conversation that happened there kept me going back. In the group I met former service members turned spouses, teachers and nurses, and other brilliant minds who had also found themselves passing their time in Korea with careers on pause. In some ways, this makes the sound of screeching career brakes easier to take. I am not an anomaly.

I wonder if each move will be like this in its own way, a feral creature to be tamed. I might lose myself a little each time, wildness flashing from behind my eyes, but then I am able to reign it in. The transitions get easier to bring back to center.

Or maybe Korea is a different beast altogether.

Surfing the floor of the subway car to keep my footing as it rocks back and forth, I look at the others who have joined this commute. The train adds and subtracts people at each stop, but the crowded car is relatively quiet. In the quiet hum of motion and energy, my mind wanders to what I'll do when I get home. At this post, Ryan maintains a brutal shift schedule.

He describes it as two on, two off, three on, two off, two on, three off. And every two weeks he transitions between days and nights. The only thing regular about it is its difficulty. They call it the Panama shift but nothing fun or tropical springs to mind when I think about it. We see each other when we can and get out when he has energy. But our grand adventure time together is not as much as I had hoped for. I am working on coping with what I want versus what we have. Tonight, I will eat alone and fall asleep waiting to hear his key in the lock tomorrow morning.

The train shudders to a stop and I step onto the platform. The streetlights are on, and the city is awake even though the sun is not up on this side of the world. I walk home, stepping on the fallen yellow fans of ginkgo leaves decorating the sidewalk, and think about how the Ten-Minute Play Festival, new friends, and dance training have helped ease the tension of waiting for what is next. It's hard to believe that a year ago I was all set in California. There I reestablished myself in a matter of months. But maybe in Seoul it will take longer, and while it makes my pulse race to consider it again, maybe monetary work won't happen at all.

I turn the key in our apartment door and head straight for the shower. I try to let my wonderings wash down the drain as I scrub away the salty sweat. Perhaps instead of asking "What will I do?" my intention needs to shift to "Who will I be?"

This time can be an unplanned sabbatical. A pause, not a stop. I turn and let the water drum on my shoulders. I know this is not a one-and-done choice. I will have to choose how I spend and, maybe more importantly, how I view each day here. I step out of the shower without answers for how to accomplish this feat, but I have the willingness to try.

~

When the ginkgo leaves have all been stripped from the trees, I face our first Korean winter. My cheeks, lips, and nose bear the brunt of the biting wind, and I brace myself for each walk to the subway. The chill seeps through my mittens. Gray clouds line the sky, week after week, and snow is on the ground. Even for a girl from Ohio, the bitter shock of the frigid temperatures rattles the bones beneath my skin.

Today I decide to give myself the gift of clinging to the inside of our apartment with books for company. Since Christmas I have been working my way back through some of my old favorites. With my extra time, reading and books have returned to my life. It is hard for me to remember that not so long ago, I didn't have (or make) time to settle in with a good story.

When the timer sounds from our cozy kitchen, I place a bookmark between the well-worn pages of *Anne of Avonlea* and leave my comfy chair. I pad across our apartment floor in thick woolen socks to check the soup simmering on the stove and the bread in the oven. Along with a return to reading, I have started to learn to cook here. In our postage-stamp-size kitchen, I have one square of counter space along with the washer and dryer to prep food on, but it suffices. I check out cookbooks from the post library and watch tutorials online. Who knew you needed to pat down the meat for proper browning? Or that pounding dough to reactivate yeast is like working a giant stress ball?

Once the meal has been tasted, seasoned, and stirred, I walk down the hall and pull my camera out of my backpack. I click through my adventures from this past week. Even with the cold weather I have still found adventure buddies who are willing to brave the temperatures in pursuit of art museums, new restaurants, and cute cafés. Each outing adds stories to the blog I started when we were in California. It began as a way to keep our family and friends apprised of our travels. I had no idea then how this hobby space would breathe life into me now.

I smile at the snaps from two of the cute cafés I recently visited. One had live sheep! The other had an entire vintage Volkswagen Bus inside to sit in. I am already envisioning the way I'll add the images, text, and directions in case anyone else stumbles across my adventures in Seoul. Maybe they will be inspired to have an adventure of their own.

This reminds me of the last conversation with my mom. She told me her friends at work look forward to my posts, and recently one friend told Mom she was "channeling her inner Lindsay," as she headed off to try something new.

A giggle erupted from deep in my chest. "Really?"

"Yes! Isn't that fun?" Mom asked.

"It really is. Thank you for sharing that with me."

"It takes courage, putting yourself out there."

All I could do in response was breathe out a small, "Hmm." Was it courage?

Courage feels like a big word for the little steps I am taking each day. Courage is a word reserved for feats of heroism, not for a military spouse trying to stay busy at an overseas post.

"It is not easy to get out," I told Mom. But each time I engage with Seoul it peels back beautiful layers. Piece by piece I'm tasting, seeing, and experiencing a culture far from my own, yet it houses human beings similar to me. I could have realized this without being stationed in Korea, but being here reinforces an intention to look for what brings us together rather than what drives us apart.

"I can't wait to see it," she said. We talked about her and Dad booking tickets for next year, before finishing our chat with how they were weathering this winter. She told me about the books she was reading and the soups she had simmering on her stove.

"It is the art of staying in," I said.

"You are also learning to stay in," she said.

"True," I said. "I am learning to be my own good company."

"Never a bad time for that," she replied.

"You have you for your whole life," I said.

I set the camera down and pick up *Anne* again. I sink into our soft couch and let the aromas from the kitchen wrap me up in delightful anticipation. Sharing this meal with Ryan later will be lovely. My eyes work to reengage with the book, but I'm still thinking. *What if the rest of the time here in Korea went like this?*

I have never been one to slow down. Like a shark needs movement to force oxygen-rich water over its gills, I have felt my existence required me to keep moving from task to task. Job to job. Purpose to purpose. But perhaps some chapters in life are meant to be a pause. A reset. In the last seven months I have begun to learn how to be myself without work and movement defining me, and it has been okay. It has been a gift.

A week or so later, I am washing dishes when my cell phone rings. I recognize the military prefix; the call is from on post. I reach for the towel to dry my hands and press the answer button.

"Hello?" I ask, a dab of soap suds running down my wrist.

"Hi, I'm calling from Child and Youth Services," the voice tells me. "We offer programs for children and youth through our SKIES Unlimited program over at Yongsan Garrison."

"Oh, hi! Great to hear from you," I say, remembering the independent contractor packet and background check form I had turned in at the beginning of summer. SKIES stands for School of Knowledge, Inspiration, Exploration, and Skills and includes all kinds of classes for military kids—classes in need of teachers.

"Are you still interested in teaching?"

I am silent. Soap bubbles drift down my forearm to my elbow.

"Hello?"

"Um, yes! Wow!" I finally manage to reply. "That would be great."

The program rep invites me to a meeting in a few days to discuss my proposed class ideas and schedule and says I can start as early as next month. When we hang up the phone, I finish drying my hands. The wind whirls outside our living room window, the lights of Seoul winking from beyond the snowflakes.

My sabbatical is over.

Worth It

"WE ARE GOING TO HAVE A DANCE RECITAL. It's coming soon so you'll need to register to be part of the fun!" I tell my students.

It has been months since I started my dance program on base. When I announce the recital date, I smile as parents plug it into their phone calendars. *This is going to be amazing,* I think.

I turn to my dancers in each class and say, "We're going to start learning our routine today! Do you want to hear your music?"

"YES!" my three- and four-year-old preschoolers answer in a scream, deafening me for a minute.

"Yeah!" my five- to ten-year-old students say as they clap their hands.

"Really?" my teens ask, looking excited and nervous.

At the beginning of the program I learned that the garrison offered a chance for students in extracurricular activities to participate in one or two public performances on post. The first event I went to was in a parking lot. My ballerinas had to try to demonstrate their routine on squashy gymnastic mats. It was not great. After that experience, I gently urged my bosses to think bigger, to have a dedicated event, a recital to showcase the students' hard work and talent. I asked for a stage, and why not? What was stopping us from not only providing stellar instruction but also giving kids a chance to share their skills with their families and friends?

They told me to be patient and build my program. Scanning my roster of classes a month ago, I saw it was full. That means

I have enough ballet, tap, and jazz classes to entertain an audience for more than an hour and a half. I proposed the recital again, gained support, and got the okay to have our recital in the high school auditorium. And for a recital, we need costumes.

I hand out a printed flyer to inform parents about costumes, hair, and makeup. "Costumes will cost five dollars. Your dancer will receive a skirt and a hairpiece. They will need to wear their black leotard and tights underneath."

"We are making the costumes ourselves," I tell Ryan as I sit on the floor of our apartment. He is on the couch while I am assembling piles of class parents' cash into envelopes to purchase the material for the costumes.

He glances at me and says, "Wow. This is a lot of work, babe."

I mark another envelope with a red marker and shrug. "Each class has a volunteer parent who is going to assemble their class's hairpieces. So I'm just on tutu duty."

"Tutu duty?" he asks, swinging his body to now face me from the couch. "How many tutus?"

"Ummm," I pause to count through my classes. "Sixty-five."

He lets out a low, slow whistle. It sounds like when Wile E. Coyote launches himself out of a cannon after the Roadrunner. A long, arched descent ending in demise.

"There was no other way to make the costume part work," I say, and it was true. I couldn't order from an American dance costume catalog and get it here in time, nor would the post office on the post allow me to use our APO for a business venture. While there might have been a way to shop for costumes at the local stores, making our own would be the most cost-effective for the families with kids enrolled in classes. And Seoul's extensive fabric markets would provide the perfect resources.

"It's going to be worth it," I say. Ryan smiles but doesn't say anything more.

I turn back to survey my envelopes, each with a dancer's name scrawled on the front. I imagine handing each child their finished costume. I already see joy.

Yes, I think. *Yes to leaping, sometimes before looking. Because if I look, I might not attempt to do the impossible.*

I stand in a fabric vendor's tiny stall. Bolts of tulle are stacked up and over on shelves all the way to the ceiling. Not one square inch of space is spared, and it's like being surrounded by soft rainbow spools of cotton candy. I shift my weight on the hard cement floor.

This adventure was fun when it started: combing the labyrinthine fabric market with a crew of dance moms for all the right colors and each special detail to add to the costumes. Up, down, around we've gone, weaving through tight spaces, searching for the perfect materials. But the fun dried up hours ago like a puddle in the summer sun. The market is like the *Alice in Wonderland* of fabric, and we've fallen down the rabbit hole. We negotiate prices with clumsy hand signals and broken Korean.

"I think she's saying she'll let us have the whole bolt for 10,000 won," Ashley, one of the dance moms, says.

"Here, use my phone again," I say and pass over my phone, open to the calculator. The woman punches in her price, and we calculate the cost in dollars.

"Great, yes, let's get it. That's more than fair," I say.

Ashley nods and opens her class envelope. We pay, and the shop owner slides the bolt into a giant trash bag. We motion at her not to close it as we scan our list of dancers and routines. We point up to more bolts. The owner swings pieces of fabric into the air, and the tulle catches an unseen breeze as she cuts the meters we need. Delicate trimmings float to the floor like downy feathers.

Our minds work overtime to get the math right. We only want to make this giant shopping journey one time. It was a headache just to coordinate schedules to get us all here together.

By the end of our trip, the bag full of tulle looks like a giant puff of clouds gathered at sunset, spilling over with pink, blue, purple, green, and yellow tufts. Sweat rolls down my face and trickles down my spine. I think there's even some pooling behind my knees. My feet ache from roaming in and out of shops and climbing stairs. But the quest has been victorious. My shoulders are weighed down by bags stuffed with yards of sequins, elastic, ribbons, and buttons.

"Is that everything?" I ask, and another mother checks our list.

"Yes! We did it," she says, and we high five all around.

We climb down into the subway like a group of *Project Runway* Santas, with black trash bags dangling from our elbows, wrists, and shoulders. Back on post, we march onward to Ashley's house. She pulls out coffee and snacks, but we all hit the water first and guzzle like we've been through the desert. Over the next three hours, we cut and bag fabric for each class, in preparation for creating the individual tutus. We will still need to sew the elastic waists, then knot and tie on each layer of the skirts by hand, which will take another few weeks of work. Our hands become stiff, and my body dreams of lying in bed, but my gratitude grows with each passing task.

I look at this group of parents who have gathered to make an event special, to create a unique experience for the kids in these classes. I am not a parent yet, but I marvel at how each one has stepped up to volunteer their time, because they have time to offer. And while we assemble the pieces, they do not begrudge those who can't take on the extra effort this time. We laugh our way through the last of the packing, and I leave thinking about how I am learning to serve where I can. There is a chance to have precious discernment about when I can volunteer for the greater good. I must ask myself if I am able to serve, do so when I have the capacity to, but also step aside when I don't. Yes, it's possible to keep changing what I do and how I work.

Months later, Ryan snaps a photo of me after the recital. My arms are laden with fresh bouquets of flowers, and my face is spread with a grin. Where I once could not imagine a life without performing myself, I have been fulfilled in a different way by helping others take their first bows. Whether a student of mine goes on to live a life as a performer or not, I hope she or he gains confidence, kindness, and an appreciation for the arts. Sixty-five tutus now rest in the homes of little dancers, and I hope they pull them out often to leap and twirl and reimagine themselves as the brave stars they were when they shone out upon the stage.

Yes, it was worth it.

~

My slippers shuffle me into the kitchen to watch the last of the coffee drip down into the pot. I fill a cup and pour in some cream before walking over to check the bread dough. With a sleepy smile, I stare down at it. Bubbles have formed at the top and it has risen so much it almost touches the tea towel I wrapped around it last night. I take a sip of coffee while I set the temperature of the oven. I note the time, six o'clock, and check my list of what we're preparing: two types of bread, honey ham, green beans, glazed carrots, mashed potatoes, gingerbread cookies, and brownies. I feel tired thinking of the work ahead of us today but I'm motivated when I look out to our living room and imagine the crowd that will gather in it later this evening. Yes, a feast for our friends.

While the oven finishes preheating I roll up my pajama sleeves and punch down the dough, reactivating the yeast. I form it into a rough ball. I oil our big, red Dutch oven and place the dough gently down inside, slicing an "X" into the top with a sharp knife. With one task done I move on to the rest. By the time Ryan gets up I've made it two hours into preparations and am lounging on the couch, having traded my coffee for water. He looks over everything and asks, "How can I help?"

I slide over so he can sit beside me. "I have the list here," I say. "I have never appreciated my mom more."

He chuckles. "Our first Christmas party, though. It's going to be appreciated. You got something for everyone."

"The ham will be the star," I say and raise my eyebrows.

Ham is hard to come by in Seoul, but we have procured a huge one from the commissary. Several of the other dishes we are making aren't common out in the city either, but that was the point of this holiday celebration. Only a few of the friends coming this evening are military. The rest are expat friends I've made through the theater community. Lacking access to the commissary, that wonderland of American products, they are all excited to share in the exotic treats of canned green bean casserole with cream of mushroom soup (the crispy fried onions!), mashed potatoes, and yes, there has been great talk of The Ham. While everyone loves the traditional Korean fare they eat on the regular, it's good for the soul to have a taste of home.

And that's what we have this Christmas, I think. *Home. Home away from home.*

Last year, Ryan worked Christmas and Christmas Eve. As the only member of his team without kids, he took the holiday shift so those with little ones could be home. I spent those days rotating amongst the homes of other friends who adopted me. I felt welcome and loved but I'm happy he's home this year and that we have a robust group of friends to share a party with. So much has shifted this year, for the better, mostly because I've shifted my mindset about what is necessary for a special holiday.

While Ryan busies himself in the kitchen with brownie batter, I place our white elephant gifts under our tree. It reminds me of the tree in *A Charlie Brown Christmas*, with its little, bare branches bending under the weight of the six ornaments I have added. It stands at just two feet tall, with one strand of white lights. I brought four ornaments from home: two of Ryan's, two of mine, and now we have added one Marine Corps ornament and our new Korean ornament, a little figure in her *hanbok* surrounded by clay *kimchi* pots that make a muffled chime when they clank together.

I always knew it didn't take all the trimmings to make a real Christmas (or Thanksgiving, or Easter, or any of our other big celebrations), but our overseas life has challenged me to believe that more deeply. It is easy to get into a scarcity mindset and lament what isn't here. Even without our immediate family, a big tree, and traditions like visiting a Christmas tree farm or driving past neighborhood lights, when Ryan slides out a pan from the cabinet and begins to pour in a layer of sticky sweet chocolate, I feel only the abundance of what we do have this year.

We have chosen to provide the gathering place this time, and I am glad to be here. This moment is becoming a memory and a tutorial on how to handle seasons that don't fit the traditional mold.

It has been an exercise in saying yes. Yes to merriment. Yes to new traditions. Yes to making the investment of time and effort in a feast for this joyous gathering with the good people we are honored to know here.

Yes to providing the sought-after ham. And at the end of the day, there is not one bite of it left over.

It is hard work. An investment in meal planning, hospitality, and friendship.

And it is worth it.

~

I wrap my arms around myself, tugging my scarf tightly around my face. It is January of my second winter in Korea, and I am still not accustomed to the bitter cold. I warm myself with the thought that our move to Hawaii is only three months away. My toes are already growing numb as I wind through the back alleys to reach the main stretch of the Itaewon district. I look up the street, with its flashing signs and rows of international restaurants, nightclubs, tailors, and coffee shops. I bury my face in my scarf and focus on walking to the end of the strip.

Kaylynn has asked to meet me for coffee, and we agree on a tiny shop in her neighborhood. I told her I didn't mind the walk, which is true, but I also wanted to ensure she did not get lost. I remember it's hard to be new, but I am leaving soon. This is a courtesy visit. As I walk, my shoulders pull inward, but not from the cold. They scrunch in with the leaden weight of shame. I do not want to come to this coffee date. I can't shake the feeling that it's a waste of time, and my rudeness feels as foreign as these streets used to be.

The street light flashes red before me, and I wait, stamping my feet to keep the blood and warmth flowing. I replay last week's conversation in my mind.

"I don't know, babe. We're about to leave," I had said to Ryan after I received Kaylynn's text invitation to a coffee date. I feel like I do not have the energy to make a new connection.

"She's nice, though, right?" he asked.

"Well, yeah. I mean, remember how I told you she was sitting behind me on the bus in China?" I said. A flicker of a smile hits my lips. Kaylynn popped right over the seat, tapped me on the shoulder, and introduced herself, a slight Texas twang in her voice.

"Right. You hung out with her on that trip. And you've seen her at other things, so what's the deal?" he asked.

I picked at my cuticles and shrugged. I did not know what my deal was. I was normally excited to meet new people and to share Korea with them. But as our months in the country wound down, I felt like I was in the middle of a bad breakup. One I didn't choose. Even though I knew moving season was coming, even though I knew we could not stay at a duty station, that knowledge did not help the way I felt. We were at the mercy of what the military needed us to do next, whether or not I had embraced our current location. My heart was divided. I wanted to push our time away and also pull it close.

The light turns green and I surge across the street, dodging a broken cobblestone as I climb up the next curb onto the sidewalk. I remember what had I asked Ryan next: "What is the point?" He straightened his back and looked at me.

"Maybe the point is that this coffee is not for you," he said. It sounded harsh, but sometimes the truth is. While I might not have time left to be Kaylynn's person in Korea, maybe it was still worth the effort to connect her to the people who would be her friends.

I risk taking off one glove to check the address of the coffee shop. My fingers curl in the cold while I swipe my phone, and I realize how close I am. One more block later, I am snug at a table inside, watching for Kaylynn. The winter wind blows her through the door, a tangle of coat, boots, hat, mittens, curly blonde hair, and a big smile. She sits down across from me and we order lattes sprinkled with cinnamon and spend the next few hours talking. The conversation is easy. We are quick to connect and have much in common. I leave the table wishing she had been here the past two years with me.

Over the next few months, I get ready to go while she settles in. Kaylynn's calendar fills with new explorations while mine starts to empty. She expands her circle of friends. I realize that my circle, because of snowy day lattes, does not contract. Now I understand it is never too late to make a friend. And it's never too late to be the friend I had wished for when I was new at a duty station.

I remember walking back to my neighborhood after meeting Kaylynn. I am pretty sure I sauntered down the street with a spring in my step. The chill was no less daunting than when I'd left our apartment, but the warmth of our afternoon together kept me humming all the way home. Making new friends takes effort, as does keeping old ones. I have had to let go of more friendships than

I like to remember because of distance and shifting circumstances. But the friends who remain, the relationships that continue to expand as time goes on, will always be worth it.

Twelve months later, when I'm settling into our new assignment in Hawaii, Kaylynn surprises me with a direct message on social media: "We're moving to Hawaii!"

Cue the next round of coffees and a deepening friendship that might never have been.

Still Learning

MY MOM'S EYES WIDEN as I hand her a stack of pink pajamas along with a towel and hair cap. I place a locker key attached to a spiral rubber bracelet around her wrist.

"We go through here," I say as I hold open the door to the women's side. "You didn't want a snack, did you?" I ask when I see her glance over the array of Korean snack cakes, hard-boiled eggs, and hand rolls.

"No, I'm good!" she says and walks through to the dressing room.

"Our lockers are right next to one another," I say and we look around for numbers twenty and twenty-one. Once we locate them, we start to undress, folding our clothes and putting our purses and shoes onto the metal shelves.

"So, you wear your pajamas in all the public spaces," I say, slipping the pink top over my head. "But when we're ready to head into the bathhouse it'll be—"

"Time for my birthday suit!" she says and chuckles, bending forward to brace herself while she slips one leg into her pink pants. We slide our feet into the sandals we brought and check ourselves in the mirror.

"Hold on," I say and pull my phone from my pajama pocket.

"You can keep your phone?" she asks.

"Only on the public level," I say. "Now pose!"

We look into the mirror, and Mom slides her arm around my

waist. She presses her cheek into mine. I relish her warmth and the familiar smell of the tropical lotion she wears. She leaves hints of coconut and lime whenever she visits. Sometimes, after she goes back home after a visit, I lie down and press my face into her pillow, just to hold onto her physical presence a little longer. The twinge of sadness hits my heart for the first time this visit. It is amazing to share Seoul with my parents, but it was hard to wait an entire year to get to see them.

I try to shake off my grief and be in the moment by lifting my phone up to frame us.

Capture this time. Be here now. Worry about the goodbye later.

Dad has already gone back to America by himself. I remember watching him enter the security line at the airport, waving at us, and I think about how he told me I was so brave, navigating this country and a new job. He told me he was proud. I never knew I could be equally proud, watching both parents summon the courage and adventurous spirit to let Ryan and me lead them on this visit and have good humor about it all.

"Now this!" Mom says and blows a kiss into the mirror.

"Ha, yes! That's good," I say and join her.

Mom and I giggle as we change pose positions and vamp for a few more selfies. We laugh harder when it occurs to us that we resemble pink flamingos in nursing scrubs.

Mom surveys the public area, which has different themed rooms. One is silver and says it has healing power, while another has crystals spread throughout and is painted a florid purple. We peek our heads into several other rooms and walk past couples, families, and groups of guys and girlfriends on the main floor.

"I'm ready," she says.

I raise my eyebrows at her with a smile. "Okay, let's do this!"

Back at our lockers, we strip down. I hand her the hair cover, which is like a soft fabric swim cap but shapeless like a hairnet.

She pulls it down over her short hair and tucks her wisps behind her ears.

"There are towels inside," I say and turn my full moon to her as I lead her back for her first *jjimjilbang* experience. When I pull open the door, a rush of hot steam hits us both, and we shuffle our sandals across the tile floor. Mom comes beside me to hold onto my elbow, keeping herself steady on the wet floor. "Are you doing okay?" I ask.

"Yeah," she says as she looks around. There are pockets of women scattered around the room. Four scrubbing tables are set in a corner, each the size of a massage table. We have decided not to pay extra for women to scrub dead skin off our bodies today. The last time I tried it was both an interesting and intense experience. The amount of skin left on the woman's scrub mitt made me wonder if she'd made it down to my muscle and bone.

Mom and I dip into one of the hot tubs with carved seats and water jets. We bubble in there for twenty minutes before moving to a slightly cooler pool, and then I challenge her to a polar bear dip into freezing water. She emits a tiny shriek before setting her lips into a thin, hard line, the Maine native in her adjusting to the harsh but refreshing water.

I twirl around in the pool, giddy we are having this moment together. I remember all the times we went camping as a family, Mom, Bethany, and I crammed into a shower stall. Sometimes we had to feed the shower coins to keep the water running. Each time we huddled together, washing and rinsing and laughing, with Mom sometimes whisper-yelling at us, "Keep your towels off the floor!"

"What next?" she asks as we climb the ladders to leave the pool.

"I brought our hair conditioner and facemasks," I say. "You grab us two stools at the mirror stations, and I'll run out to the locker room to get our supplies." I gesture to the two lines of wash stations and she nods. When I return I see her on a stool, casually looking around the room. Her stool sits in front of her own personal mirror and shower head. I take the one to her right.

"Pretty neat, isn't it?" I ask. I show her how to turn the handheld washer on and start working my way through my hair.

"Yes, it's been fine," she says, tilting her head into a warm stream of water as she smiles. "I like to see all the families in here."

I start to work a dab of conditioner through my hair. "Remember when my friend Christie visited?" Mom nods. "She said the same thing. She wondered how women would feel about their bodies in America if more than just the big cities had bathhouses."

"Yes, knowing how bodies change over generations," Mom says.

"I like to think I had that," I say. Mom takes the conditioner bottle from me and meets my eyes in my mirror.

"Really?" she asks.

"Yeah, I mean, you and Nana never made it weird to change in front of one another. You just never acted like it was a big deal."

I remembered us all going to pick out a few new pants and tops each fall before school started. It was fun to try on clothing together, and until this moment, I hadn't stopped to think about how tough it is for a mom to get her young girls through adolescence. Maybe I didn't have to think about it because I just knew Mom and Nana would be there for me to select bras, undies, pants, and tops.

I catch Mom's face in the mirror. She is working to apply the paper face mask I bought for each of us. When I put mine on we look at one another and laugh.

"Boo!" she says.

"They are kinda scary," I reply as I study the tissue-thin white sheet stretched across my face. We sit and shave our legs as the facemasks work their magic. We peel them off and get ready to lather the rest of our bodies clean when we hear a rise of giggles to our right. We peek over to see what looks like a mom, grandma, and child around age six washing together. The grandma is sitting on the stool as the mom and daughter scrub her back. Mom looks back at me.

"Do you want to wash my back?" she asks. I do.

She sits regally up on her stool, and I add soap to my pink loofah. I lather it and swirl slow circles across her shoulders and down her arms, then trace the notches of her spine. I pause with the suds dripping down my hands. I bend forward to kiss the top of her head, and that faint whiff of coconut lime lingers as I reach for the shower head. I guide the warm water down her back so all the suds drip down into the drain. "Feel better?" I ask, and she nods.

I sit down on my stool and cradle the loofah in my hands. It feels like a moment is caught in my throat. It holds fast there as I reach for the right words.

Moving to Seoul has been challenging, and yet it has forced Ryan and me to deepen our own little family in new ways. It's forced me to grow. But having Mom, Dad, and friends come to visit feels like oxygen to my body. They fill me up and refresh me. They remind me where I came from. I manage to swallow down the lump. It isn't caused by unease but by gratitude. I accept the paradox that I can embrace and enjoy where I am in life and still miss my parents desperately. I know we'll see one another again at our next home, but it never gets easier to rotate through hello and goodbye.

"What do you think?" Mom asks. "Do you want a turn?"

My eyes burn and I lift my face up to her and smile. She scrubs me with the same care I imagine she had for me when I was a small baby, then a toddler, and then a girl. And here we are as women, still learning about each other and the world around us.

"I haven't washed this back in a long time," she says and bends to kiss my cheek.

"Yes," I say and I lean into her kiss, allowing myself to be little and big all at once.

Question, Answer

"WOULD YOU RATHER BE PREGNANT ALONE or have the baby alone?" Ryan asks. I dig my pale white toes next to his in the sand. We landed in Hawaii two weeks ago, and the trade winds are fierce today, whipping the waves into an unwelcoming froth. He reaches for the coffee tumbler we're sharing, and I place it in his hand. He sips, then passes it back. My fingertips graze his, but we keep our eyes focused on the ocean. Maybe it has the answers we seek.

I shift my feet over his and lean my body into his side. I break my ocean hypnosis to look directly at his clean-shaven face, his tight haircut and his strong arms.

This is not a normal question you ask your wife.

My eyes beg for him to meet mine, but he keeps them on the waves, his jaw set.

Reveille blasts from a nearby speaker. The base owns this portion of the beach, so the military touches us, even here. The bugle slices through the air, and ready or not, the day has begun. My husband digs our cup into the sand, rises, and faces the sound of reveille as he stands at attention. Even in his civilian clothing, he acts as though he is in formation with other US Marines. And he is. There is no respite from this commitment.

I dust off my shorts and stand next to him, although I lean on one hip, slouch a little, and shade my eyes. When it ends, the spell of us on the beach alone together is broken. I can see his mind sorting out what orders he needs to carry out today. He snatches the cup from the sand and starts to head back up to the car.

I catch his hand as he walks away, pulling me along. I pull back, stopping him to give my answer.

"I'd rather be pregnant alone," I say.

I get to hold onto Ryan for ten months on the island. We find a cozy "ohana house," a rental cottage behind the home of an older couple, and we begin the adventure of unboxing another home and other adventures like paddleboarding, hiking, and creating another new community of friends.

On the Friday before Mother's Day, it's time for him to leave. We stand at the bus and I hold him, breathing in his scent, soaking him in. I can feel him tense. His body and mind have already left.

"You're going to be okay," he breathes into my hair. "You've done this before." I bury my face farther into him, wishing my weight could anchor him here.

I clear the lump from my throat and reply, "Yeah. I will. I'll—I'll let you know, okay?"

He pulls back from me and smiles.

"Could be—this time," he says. He was the one who bought the last round of pregnancy tests. I left them in their bag on the bathroom counter.

"Yeah. Could be." I kiss him one last time and force myself to release him.

I grip the steering wheel to drive home, but I have to pause before backing out. I let the tears go. I let my heart feel the new ache. Unlike our goodbye, the pain of deployment will not be fast and over. I let it wash over me, the fresh waves of want, the months of longing ahead. Seven this time.

Four days later, I stand in the bathroom at five-thirty in the morning. The pregnancy test is nestled on a little piece of toilet paper, my first morning urine soaking into it, a Hail Mary pass.

We can always try again when he gets back. But wouldn't it be fantastic—if—

I glance down at the test. I pick it up to be sure.

A plus sign.

I set the test down and look at myself in the mirror. I tilt my body to the side and run a hand over my belly, checking for change. I tilt the other way. My breasts feel slightly heavier, but maybe I just want them to. As I take stock of my body, I can see the hand towel behind me in the mirror. It hangs rumpled and wet from drying my hands earlier. There is a dried whirl of white and blue in the sink—toothpaste from last night. Three hair ties and blonde bobby pins are scattered across the bathroom countertop. Everything is as it was yesterday, except for inside of me. I pick up the test again and meet my eyes in the mirror. They are frozen wide.

You're going to be a mother.

I press my palm to my belly, and my eyes crinkle into a grin. Hello, you! Hello!

I swipe the bobby pins and hair ties aside, rip off a clean piece of toilet paper, and place the test in a new little white nest of bathroom tissue. I can't throw away this proof.

I walk into our bedroom and grab my phone. I lift it up to call, to text, but I realize he's still en route. I lie down, gripping the phone. I flip to my email but then close it. We did not get this far in our preparation. How do I tell Ryan? Shouldn't he be the first to know? It could still be a few days before I can reach him.

I tap in another number and lift the phone to my ear.

"Hello?" Katie, my best friend since college, answers. While she wasn't able to come to Korea, we managed to talk every week, and she has already bought tickets to visit me in Hawaii to help break up the deployment time. I hear a siren go off in the distance, cars honking, people talking. It's early in Hawaii, but New York City has already been up for hours.

"Hi. I'm—I think I'm pregnant," I say.

"Wait—what? I couldn't hear that. Can you say it again?" she says, and I can hear her breath, rushed. I imagine her crossing a street.

"I'm pregnant!" I shout this time. I imagine her feet stopping on the sidewalk.

"Ommmygggosh, Lindsay!" she says.

"I know! I mean, I haven't gone to the doctor or anything but ..." I trail off and realize maybe I'm excited too soon.

"Okay. Right. But—I'm sure it's something. Let's celebrate anyway. Okay?"

"Okay," I say.

"Let's just dance up and down for a minute," she says. I get out of bed. I imagine holding her hands on that sidewalk and she counts down for us: "In three, two, one, go!" We both hoot and dance in place, like we did in college when we got something we had been hoping for. I let my joy for this new life pulse around me.

I'm in the bathroom two days later, and when I wipe, there is blood on the toilet paper. It fans out in the shape of a rose petal, like the ones from our wedding. I still have not heard from Ryan. I dial the advice line at my ob-gyn. A nurse named Delia answers. She sounds kind. I tell her about the blood. I tell her the pregnancy test on my bathroom countertop is positive. She reminds me that my bloodwork is scheduled for Friday, and today is Wednesday.

"It could be implantation," she says. I ask her what that means. She explains briefly how my uterus is already changing and tells me, "It's best not to worry. If you don't soak a pad, we'll just see you Friday."

"And what if it's a miscarriage?" I ask.

"There's not anything you or I can do, hon," she answers. I hang up and crawl onto the couch. I hold my phone in my hand, willing Ryan to call me. My mom calls instead.

I stare down at her name and consider my options. I want to

tell Ryan our news, whatever it is, and I want to share this moment with him. But I can't reach him right now. My parents will arrive on the island tomorrow, and if I'm miscarrying—they need to know what they are about to come into. I call them back, and we share a short and emotional phone conversation. I float an equally short and emotional email out to Ryan.

As I wash my face for bed, I hear my phone vibrate. It figures. This is the one time I've not been within arm's reach of it all day. I run into our bedroom and snatch it. The number shows up as Unknown. I sigh with relief.

"Ryan?" I say.

"Yeah, babe. Hi, how are you? We're just getting a break here to check in."

I wonder the best way to tell him, but I know he doesn't have long, so I blurt out, "I'm pregnant!"

"That's amazing news," he says. "How are you feeling?"

I tell him about the blood and the nurse, about my parents and the test scheduled in two days. He tells me not to worry.

"I love you," he says, and then he's gone.

It isn't what I pictured in my head, revealing the pregnancy. I imagined a cute moment with an ultrasound, his arms around me, telling our parents together. I go to bed with my arm tucked around my stomach, willing the human within to stay.

My mom sits with me at my first appointment. We go into the examining room together. The technician explains she will have to do a transvaginal ultrasound because I'm not far enough along for the standard goo-on-tummy ultrasound. She turns the screen toward her, and the hunt begins.

She keeps her face blank. I reach for my mom's hand. It's warm and rough. Mine is cold and clammy. Mom rubs her thumb along my thumb, a comforting gesture she does absentmindedly, but I remember it from childhood. I suck in my bottom lip and bite down.

"There. There—want to see?" The tech breaks into a smile. "You have a little gummy bear baby right now." She turns the screen and I see it: a small square with nubbin arms and legs and the flicker of a heartbeat. I breathe out and look up at my mom. There's a tear on her cheek. I reach up to find mine wet too.

At the twenty-week ultrasound I am alone in a dim room. The door opens and the doctor enters.

"So if the baby plays along with us today, we'll know the gender," he says. "Do you want to know?"

Ryan and I have devised a plan.

"We do want to know," I say. "I have my phone here, so can you text my husband first?" I don't have to explain to the doctor that I've been to the rest of the appointments myself, that my husband is deployed. He knows. Since my mom and dad's visit, it has always been me. Alone. Like I wanted, I suppose?

The doctor grins. "Sure. That's a nice idea."

I walk around for three hours with our baby's gender on my phone. I buy a smoothie and do some work at a café. I have to push the phone to the bottom of my purse. Now, I'm walking around a department store when, finally, it rings.

"Hi," I breathe.

"Hey," he says. He sounds tired. "You ready?" he asks.

"Of course!" I say.

"It's a girl," he says.

"Are you sure?" I look at the text. Girl. "Is it still Evelyn?" I ask. We picked out girl and boy names already.

"I still like Evelyn," he says, "so, yes."

"Okay!" I say. I wait for more. I wait for confetti to fall from the ceiling or nearby shoppers to realize my big life moment. But it is just us on the phone together, connected and apart.

"Well, hon, I gotta head out. Love you. Glad we did this," he says.

“Me too. Bye, babe.” I turn and head to the girls department of the store. I don’t get to have a big public gender reveal moment, but I get to have my girl. I’m buying her something to celebrate.

Two months later, I walk the aisles of Babies“R”Us and survey the confusing array of gadgets. I stroke my protruding belly and pick up one of the many pacifier options. I only have a month left before Ryan gets home. My mother-in-law, who has come to Hawaii to help me register for baby gifts and set up the nursery, hands me the scanner and it beeps several times as I grab one of everything. Nook and Mam and Dr. Brown slide under the red laser beam.

We pass the strollers and I shake my head. “We’re going to do that together, Mom. When he gets back.” She nods then heads to the car seats.

“You might want to get a car seat at least. That way if—”

“No, Mom. No!” I toss the scanner onto a shelf. I want to throw it against the white brick wall and watch it shatter next to all the baby socks and teething toys.

I turn and waddle away. I’m a giant beach ball. I can’t hide. I stumble into a happy couple registering for all their newborn gear. They glance at me sideways and hustle out of the way while I release sobs into a baby swing.

I thought I could do all of this without him. I can do it without him. I am doing it without him. But I don’t want to. I feel like a child with a child within me, pouting and crying, desperate to control something, anything.

My mother-in-law finds me and rubs my back. She doesn’t breathe a word. She simply circles her hand around and around until I sniffle and turn to her with bleary eyes. I want to believe I am all alone in this. But here she is. She flew thousands of miles to be here with me, to make meals while I lie like a sick queen on our couch, to help set up our registry and turn the spare bedroom into a darling nursery. She can’t be him, but she is here.

I take the scanner from the shelf and we start again.

The deployment stretches as I do. When I'm thirty-six weeks into growing a human, Ryan returns. He strides out of the airport terminal with day-old scruff and a blue T-shirt. Not all homecomings are glamorous with handmade signs, photographers, and fanfare; sometimes it's just a gray baggage claim and a lot of suppressed adrenaline. The purple dress I'm in is light but I'm still sweaty. The curl in my hair is already flat from humidity. He gathers me in his arms the best he can, but the belly bumps into him from every angle.

"Wow," is all he can manage. When he left, I looked like me. Now we don't have many days left as just the two of us.

In the dark of our bedroom that night I can hear his silent breathing. He lies flat on his back, his ankles and arms crossed. I know he'll be like this for a month or two—we call it the "surfboard pose," and it usually happens for a while after deployment. His whole body is contained in one long line. It comes from having to sleep whenever and wherever he can. I stroke his rough cheek. I know we are both different from our time apart.

I turn and curl my legs around my body pillow and feel her. I grin. My daughter and I had lain together every night in this empty bed while he was gone. She grew out into his hollow space beside me. When she started kicking, I was not so alone in the dark. He rustles in the bed next to us. I place my hand on her and whisper in the night, "Daddy's home."

A month later, I am folding freshly washed onesies when I hear the screen door open downstairs. "I'm up here!" I call. My belly grazes the drawers in front of me as I work to arrange the pile of tiny clothing inside the dresser. I hear his feet on the stairs, and soon he is kissing the back of my neck. I lean into him. I'm due any day.

"Hey," he says, and I feel it in one word.

"What?" I say. My hands pull in another onesie to fold. This one has a pale pink tutu attached. He grips my shoulders and turns me toward him.

"We're heading out again in four—"

"Four what?" My heart begins to pound. I look up at his eyes and in them, I see sadness.

"Four months," he says.

"For how long?" I ask.

"Six, maybe seven months again," he says.

I shrug his hands off my shoulders and turn away. I look down and find the pink tulle is damp in my hand. The delicate fabric is crumpled. I toss it to the top of the dresser.

"I'm going for a walk," I say.

"Do you want company?" He asks.

"No."

I walk the streets of our neighborhood and feel nothing and everything. My thoughts counter each other.

Why did I get my hopes up? Of course he's leaving again. The Marine Corps doesn't care what we've got going on. This is how it is. At least he'll be here for the birth. What am I going to do? I will figure it out.

I think of every holiday in the coming year. All of Evelyn's firsts. I close my eyes and imagine our family. I picture Evelyn in my mind, in the pink tutu onesie, and me holding her. I move Ryan out of the frame. I push him to the edge of the photo and then I push him out of it completely.

~

Evelyn nuzzles her fuzzy head under my chin. At four months old, she has retained soft brown wisps of hair that stick up like a miniature rooster. Tonight we watch Daddy pack his bags, each one stuffed with gear. More items wait to go in, spewed across the floor. Goodbye vomit in shades of camo.

I hear my Dad clear his throat behind me. "We're stepping out for a walk," he says.

I nod at him. He and Mom have flown out to support us, which is both good and hard. Their presence keeps Ryan and me from arguing with one another as we've done in the past before a separation, but then it also feels like we are hiding from this hardship.

We are playing at normalcy. We're pretending it is all okay. I watch Mom's eyes sweep over Ryan's preparation, and I think I see a mask of hidden concern and resolve to be who she thinks she needs to be for me. The metal screen thuds dully behind them. I see them grab hands, hear their footsteps fade down the stairs.

I step across a stack of green T-shirts and boot socks and cradle Evelyn. Ryan's arm stretches and stuffs items into a backpack. His brow furrows as he examines two pairs of boots, trying to decide which one to take. He picks both.

We stay up too late getting everything together. When the time to sleep comes, I lie there in exhausted alertness. I don't want to miss a moment of him. I reach for his hand while he sleeps.

We are like a puzzle trying to fit back together, but now with too much space between the pieces again.

Pregnant alone or baby alone? Now I get to do both.

When I say goodbye to Ryan, time seems to stretch indefinitely and then snap me like a released rubber band. I notice the palm trees whisper above us with the breeze, gentle and unknowing that lives beneath them are changing. The sounds of other spouses crying, kids telling their daddies and mommies they love them, young Marines joking and laughing with one another. The underlying rumble of the diesel bus engine churns around us, turning my stomach. It's a cocktail of excitement and trepidation, shaken and poured. I realize it has been a year since I stood in this same spot, wanting Ryan to stay and hoping for our baby. Now I'm holding our baby and wishing him away.

I hug Ryan, Evelyn snug between us, her fist in her mouth, working at her pink muslin blanket. When I pull away, my dad hugs Ryan and pats him on the back. My mom looks stoic as she hugs him, but as she pulls back I see a faint mist in her eyes. I want to hug Ryan again, but they're called onto the bus. I can't remember watching it leave. I do remember walking to the car, fastening Evelyn into her car seat, and sliding behind the wheel. I freeze there, in the quiet of the car, my hands cold and shaky on the wheel.

"Do you want me to drive?" Dad asks.

"No, I think it's better that I do," I say, channeling Ryan. He does what needs to be done. I can too.

We make it home and I place Evelyn down for a nap. The house aches with little remnants of items Ryan did not pack. Pieces of him are scattered everywhere. I begin to collect them, and Mom helps me. We work quietly alongside one another. She's the first to break the silence. "Do you still want to try the beach today?" I can't think of a reason not to.

As we drag all our items to the beach, my heart gives a small swell. Since dropping off Ryan it feels as if I've been made of lead. Dragging through each next thing, closed off, better to not feel than to feel anything brewing below. But as my feet hit the sand and the salt air whips through my hair, I feel something. What is it?

I take a deep breath and walk Evelyn down to the surf. She wriggles and I hold her close while lowering her chubby feet into the water. I wonder what she thinks of all the new sensations. Mom and Dad snap photos, and I am already imagining Mom's future scrapbook page to commemorate this moment, complete with a caption that reads, "Evelyn meets the ocean at last!"

All these months in Hawaii when we've been to the beach, she's sat on my lap at the shore. I haven't taken her down to greet the sea up close until now. I wonder why I kept it from her for this long. She gurgles as the water washes over her toes. I feel my cheeks rising, a smile spilling across my face. When we drive home, my damp hair whips around me and I chatter about all sorts of things while Dad listens and Mom plays with Evelyn in the backseat. As we turn on the street toward home, I think, *Maybe I get to be okay.*

My parents encourage me to go to a hot yoga class in the evening. I start to tell them no—but realize this is another big first step I can take during this deployment. I can trust Evelyn with my parents and nourish myself on the mat. The class brings me joy and helps all of my systems relax. When I get home, I bounce back up the steps to the apartment, where my parents are resting on the couch, watching a movie.

"Did she go down alright?" I ask, sliding out of my sandals.

"She did, but she was pretty upset," Dad admits.

My brow creases. "Oh, I'm sorry about that."

"We handled it," he says.

I swallow. *Should I have been here? Was Evelyn worried? Could she feel the shift—that Daddy is gone from here?*

I head upstairs to get a shower. I'm eating cereal in my pajamas when Evelyn screams. My spoon clatters in the bowl and I cross the living room in quick steps and into her room. It's unlike any scream I've heard her emit. Not hungry or overstimulated or wet, but guttural, like she is in pain. I steady myself with a breath as I reach to bring her to me; her eyes look at me, wide but unseeing. Her screams resound through me like a siren, and all I want is to put the emergency out.

"What is it, my girl? What is it?" I whisper at first, bouncing and shushing her. She doesn't respond to me, and I lay her on the bed, fumbling with her swaddle. "Is this bothering you?" When I release the wrap she cries harder, louder, and my hands move up and down her body, checking for anything, everything. I start to cry with her, peeling off her pajamas and my shirt so that we can rest, skin to skin.

I start to call for Ryan on instinct. We have relied on one another to sort through any baby issue together so far. But his name catches on my lips and I heave another sob.

I'm it, I'm it, I'm it. I have to be the one who sorts this.

It is only when she feels her body against mine, her ear on my heart, that she starts to move from a five-alarm-fire wail to a warning cry, and then a whimper, and then a snuffle. My cheeks are hot with tears and my body is boiling over. It takes time, but I rock and soothe the both of us. I kiss her and hug her to me one last time before I lay her back down. When I come downstairs, my parents turn to watch me as I sink down onto the bottom step and heave a rattling sob.

They tell me I'll need to continue to be strong for her. I nod but cry harder. What falls from my eyes is grief. I feel the heavy responsibility to carry the weight of this precious girl forward without him. And I feel shame—that I'm somehow not strong enough already. When will I be strong enough?

Away from Loneliness

I SHIFT FROM PROPPING UP MY HEAD on my hand to resting it flat on my arm. I'm on my side, watching Evelyn practice her new trick. At six months old, she can roll across the floor and push herself up on her elbows to see (or destroy) whatever she wants. I have no idea how she is still awake and full of energy after thirty-two hours of traveling worldwide, but she is. The timer goes off in the kitchen and I heft myself up from the ground.

"Be right back, baby girl," I say to her as she rolls past me to a pile of board books.

The packaged ramen plops into my bowl, followed by a plume of steam. This sad pile of noodles cannot hold up to the fresh bowls of true Japanese ramen and sushi rolls I had enjoyed with Ryan on our trip, but it meets the need in spite of an empty fridge. After squashing up an avocado and banana for Evelyn, I capture my rolling queen and secure her into the highchair. She grabs her food and mashes it into her mouth.

"Good, huh?" I ask and tickle her under her chin. I hear my phone buzz from the diaper bag sitting on the table and dig it out. "It's Daddy!" I tell her.

"Hi, babe," I say as I open Facetime. "Wow. Is that what my face looks like?" I push at the tender bags under my eyes and smooth the greasy flyaway hairs on my head.

"You look fine. You girls get home all right then, I see?"

"We did. Sorry, I forgot to text you. I'm exhausted," I say.

“I wish I could be there to help you,” he says.

“Me too. I can’t believe it is already over,” I say and prop him up on the table while I bend over my hot bowl. I feel like I planned the trip to see him for months, and in four days, it is already over.

“I know. But we are midway through this deployment. Only three months to go,” he says. I nod while slurping up a noodle. “Listen, I can’t be on long. I ducked out during lunch so I could check on you,” he says.

I hold up my spoon to the camera. “It’s okay,” I say. “I am trying to stretch us both to eight o’clock and then I plan on collapsing.”

“You can do it,” he says.

I smile. “I know I can,” I answer.

We hang up and Evelyn blows a wet raspberry packed with green avocado-banana goo from her lips. It sprays across her tray and I sigh. It is hard to be back to solo parenting, but I am buoyed by our trip to meet Ryan in Okinawa, Japan. My first flight with a baby and I chose to go the seven and a half hours from Honolulu to Tokyo, endure a six-hour layover, and then another three-hour flight to Oki. I left thinking I would carry the weight of the journey alone, but we met airport angels along the way.

I sat next to two teenage boys on the way to Tokyo. At first they looked horrified but then ended up playing with Evelyn for a good portion of the flight. Then there was the sweet grandmother from Kansas I met getting off the plane for the layover. She had a broken foot and was on her way to see her son in the navy, and I had a baby in a stroller. We decided to join forces and slowly made our way to the food court. She watched Evelyn while I ordered and carried our food to the table.

Countless others saw the baby strapped to my chest and hoisted suitcases up into the overhead bin or onto the security belt. Someone peeled me out of my jacket when I came upon an unanticipated security line and was wearing Evelyn. Amazing.

I arrived in Japan dripping sweat and fatigue, and I returned home much the same. But my heart is different. I am weary, but it was worth it to wrap our little family in Ryan's embrace. I wipe down Evelyn's face and her tray, kissing her cheeks as I carry her to the bathtub upstairs. I have angels here in Hawaii too. In the last three months I've had a visit from my parents. Katie came all the way from New York and spent a week lavishing Evelyn and me with hugs and attention.

I have Kaylynn on the island now, too, with her family. She and I go to a mom-and-baby fitness class at least three times a week, and then we carry our babies to the beach to play. Sometimes we go back to one or the other's houses and bathe the kids together, putting them down for naps. Then we sip coffee and decompress. That first coffee date in Korea has led to months of support here in Hawaii. Kaylynn has become my military spouse sister, helping me laugh and cry my way through the tough days without Ryan. I feel like Evelyn and I have what we need for the next few months.

I turn the lights off downstairs and darkness invades the room, making me shudder. If I am honest, the nights and weekends are the worst. This is when the quiet of the house looms over us, loneliness sinks into my bones, and the deployment stretches before us, leaving me with an ache for Ryan that threatens to pull me under.

After Evelyn is scrubbed and nestled into sweet cream footie pajamas flecked with stars and moons, I shower and wrap myself in my softest sleep shirt. I lie in bed, feeling an overtired jitter sizzle up my spine to throb at my temples. I reach for my phone and flip onto my side. I open an airline website and start to plug in locations away from Hawaii. My mind starts to arrange trips and people into place. Evelyn and I already made it all the way to Japan. We have passports. Where else can we go?

By morning I have a plan. Hawaii to Canada for the wedding of my friend Veronica, whom I met in Korea, then to Ohio to see

Ryan's parents. From there, my parents will pick us up for a trek to Arkansas. I'll see my sister in Dallas, then make it back to Hawaii. I run it all by Ryan, and while he thinks I'm a tad crazy for toting our child across North America, he understands. I told him that Evelyn and I can stay here in Hawaii, and it would be fine. But what if we could be more than fine?

"I can make the rest of this time an adventure," I tell him. "I can do this."

~

"I can't do this," I cry into the phone as I mop up the liquid poop that streaks across the bathroom floor of Veronica's parents' house.

"Yes, you can do this, Lindsay," says Mom, her voice coming in firm and clear from Arkansas to Canada.

"It's all over her and me," I say, eyeing the culprit, who is gurgling with glee in my arms.

"Is the dress ruined?"

"Probably not." I look down at the stains across my lavender bridesmaid's gown.

"Go get cleaned up and call me back," Mom orders.

"All right," I say and suck in another cry. I hang up and take Evelyn to the bathroom, where I rinse both our dresses the best I can. I hold her with me under the lukewarm shower spout and she laughs. I can't laugh with her yet. I change us into something wearable, throw the soiled garments into the washing machine, and grab my purse. I call Mom back.

"We're clean. And we're going," I say. At least it happened before the reception, not before the wedding.

"Good," she says. "Stay for the meal, maybe one dance, and then come back," she says.

"Thanks, Mom." I appreciate her cheering us onward.

On the drive to the reception, the Saskatchewan prairie spreads out for miles. The last of the sun's light flickers across fields of wheat, illuminating the crops so that they ripple like a yellow ocean in the breeze.

Evelyn and I arrived ten days before the wedding to help with the final preparations for the festivities and get some one-on-one time with Veronica. Seeing her radiate with happiness on her wedding day, having this chance to reunite years after our time in Korea, has been worth every tired day.

I shake my head and look into the rearview to see Evelyn's toes kicking in the air. We are in month four of the deployment, and I am discovering I can only outrun the grief of missing Ryan for moments at a time. It has caught up with me in this unfortunate moment, and it is kicking me from behind. I pull into the gravel parking lot of the reception hall, put the car in park, gather Evelyn in her car seat, and enter the party.

The strength I have felt coming to Canada does not have to dissipate because of one poorly timed baby poop. But as I watch families hug and couples nuzzle in to kiss and dance, I recognize an ugly emotion bubbling to the surface.

Hello, envy.

This ugly emotion threatens the edges of this adventure. I find my seat and place the car seat in front of me so I can rock Evelyn with my foot. Veronica's sister brings me a glass of white wine. She sets her hand gently on mine. She says something like, "I think what you are doing is really brave," and I bite my lip.

Right now, I am not brave. Right now, I'm mad that I am alone and angry that my child has pooped all over us. I miss Ryan and I don't want to be here without him.

But perhaps that is what makes me brave: I am here anyway.

I thank her and sip my wine. While rocking Evelyn's seat to keep her calm, I manage to eat dinner. We leave shortly after the toasts. Veronica winks at me as I exit. Evelyn fusses that night from her portable crib, and I pull her into the bed with me.

As her small body presses into mine, I think about how Evelyn and I must keep living our own stories, even without Daddy appearing on our daily page. And maybe part of life is being brave enough to allow all my emotions to exist. To let the wretched fury and envy flash through, to trade the guilt I feel about them for acceptance. In the morning I pack the car for the next leg of our journey and drive us through the golden prairie to the airport.

~

The brakes squeak and exhaust huffs out the back of the bus as it pulls into the battalion parking lot. I maneuver to get as close to the entrance as possible. Marines pour through the doors, looking for and making their way to their families. Ryan is in the middle of the bunch. He steps off the bus in a blue polo and gray slacks. He is as handsome as I remember, and I grin while holding Evelyn up in front of me. I watch his eyes widen as he takes her in.

"She's so big," he says softly when he gets to us. He reaches for her but her body wriggles back toward me and he drops his hand.

"She will be fine, but let's give her a minute. Over here," I motion with my head and we find the grassy spot in the shade of a tree I had scoped out earlier. I shimmy the diaper bag from my shoulders. "I brought a picnic blanket and some toys. I packed a small picnic too. Are you hungry?"

Ryan nods. I'm sure he's tired from the flight, but we discussed how it might be better for Evelyn to have some time with Ryan before heading straight back home.

"Look at her! She sits on her own already?" he asks but shakes his head, "I knew that, it's just—"

"Seeing it is different in person, I bet," I say and he nods. I hand him a ham and cheese sandwich and a pack of baby crackers.

"She really loves these. See if she wants one," I say, and we repeat this process with a baby food pouch and different toys. I watch the two of them carefully. I want them to connect quickly, for both of their hearts' sake. We had friends and family tell us reunification would not be a big deal.

"She's a baby, and babies adjust," they say, and she has. She has adjusted to fourteen airline flights, three different countries, multiple bedrooms, and numerous people during this first year of her life. But now she has to adjust to having a mom and a dad again. What does it mean to be this family of three? We have to adjust too.

By the end of the picnic, Evelyn allows Ryan to carry her and buckle her into our car. At home we continue to slowly integrate ourselves. I watch as Ryan naturally brings his element of fun to the scene, teasing her with the laces on his slippers. He shuffles backward across the floor and she chases after him, crawling to catch the dangling leather strands. He is a patient observer as I move us through dinner and our bedtime routine. He sits next to me in the bathroom as I bathe her. I watch as he scoops water with his palm and trickles it down her bare back. Every motion he makes is with caution and gentle care.

"Do you want to read her a story?" I ask. He shakes his head no.

"For the next little bit, I want to keep watching," he says, and I tell him that is fine. I appreciate that he wants to slide in alongside us at home, that he is not intruding or forcing the situation. We both kiss her goodnight and after I lay her down, we hold hands and leave the room.

Ryan leads me downstairs and we flop down on the couch. He stretches his legs out long and I snuggle in beside him, finding my favorite nook and resting my head on his chest. There is much to say, but words stay lodged at the back of my throat. I wonder if tonight it would be better to leave them unsaid.

Observing Ryan and Evelyn today has shaken my understanding. I know I waded through a myriad of emotions being without Ryan for the past six months, but tonight I acknowledge a new one: embarrassment. I am embarrassed to think I did not focus on what might have been Ryan's experience over the past six months.

It was hard to be the only parent, the one left behind. But what was it like to be the one out on mission, the one who missed making memories with our child?

I bury my head in Ryan's chest and concentrate on the rhythmic thumping of his heartbeat. This is not a night to close the gaps that have formed in the past six months. It is a night to simply be in the presence of one another again, a presence that contains, for better or worse, our triumphs and our failings from this deployment.

We will have to discover ways to work through the life we lived apart and the one we now decide to live together.

Marching Elephants

WHEN I WAKE UP, I SEE the old elephant mobile, sparkling in the morning light of my parents' basement. Evelyn's cries pull me from my covers. I roll off the twin cot, into the closet-turned-nursery. I brush past my mom's hanging sweater bag to turn off the noise machine. Evelyn's fuzzy toddler head is already peeping over the Pack 'n Play, waiting for me.

She lights up like a beacon when my face appears and I smile back at her. As we head out of the closet for a diaper change we pause by the elephants. It has become our morning ritual over these last two months to stop and brush the mobile and its jingling bells. We begin our day with this sweet music.

Her chubby fingers reach up and swipe at the mobile. A belly laugh rises from her and helps me chuckle. She pokes one of the bright beads with a fingertip.

Mom has picture frames lining the bureau in this room, and I carry Evelyn over to look at them.

"See? See Daddy and Mommy?" I point to the photos of Ryan and me, our fresh teenage faces beaming back at us. While some of our teenage stories are held in the ancient pages of my journal, some are stories we share repeatedly. I bounce Evelyn as I start to recall the memory of the marching elephants.

"I cannot believe we will be there at the same time!" I had said to Ryan over the phone, as I flopped down on my bed. I ran my hands over my new black satin comforter. In my senior year of high school, Mom helped me transition my bedroom decor from little kid to big kid. We added a cheetah-print throw and bright red accent pillows, but I saw remnants of my childhood when I gazed up at the ceiling. My eyes followed the blue and pink floral border around the room as I listened to Ryan's voice.

"Yeah, it's pretty cool. We'll have one day," he said.

"One awesome day," I said.

Awesome, because Ryan and I discovered we'd both be at Walt Disney World in Florida on the same day for different reasons. I'd be performing with my drill team and marching band. Ryan's Disney trip with his high school show choir would overlap mine by one day.

Besides getting to see Ryan, I'd also get to tick off an item from my bucket list: dance at Disney World. That year the list had grown substantially, with countries to see and more places I wanted to dance. Disney World was at the top. This accomplishment paled only slightly in comparison to the unique timing of the trip—and getting to see Ryan.

I rolled out of bed and back into the mess on my floor.

"What will you be wearing?" I asked as I started to throw different outfit combinations together next to my suitcase.

"I'll surprise you," is all the hint he gave me. We talked in circles of sweet nothings as we finished packing together over the phone.

"See you in a week," I said and we hung up.

Seven days later I stood with a fellow drill team dancer, Casey, at the Mexican pavilion at EPCOT. "He should be right here," I said. She pulled her sunglasses over her eyes and squinted at the building, a giant remake of an Aztec temple. I scanned the area and took a calming breath. Ryan should be here. We waited another ten minutes and Casey shifted from foot to foot.

"There! Right there, see?" A tall man-boy strode in front of the temple, wearing a collared black short-sleeve shirt and a white

Panama hat pulled down over one eye. "Aww, quick, hand me your camera!" Casey clutched at my bag with her long lacquered nails. I tossed her my disposable camera, scrolling to wind the film before I sprinted off to greet him. When I developed the film later, I'm a blonde blur running to his outstretched arms.

"Hey," he breathed into the top of my head.

I beamed upward into his scruffy chin. "Hi!"

Casey caught up to us and Ryan lifted her off the ground too. He has always had the skill of making my friends feel seen. After our joyous greetings, Casey unfolded the map.

"Let's go to the pagoda first. I want to see the acrobats, then we'll make our way around?" she said. We agreed to the plan and set off. My feet barely touched the ground. Ryan felt like a cloud, lifting me up. My heart squeezed in giddy love. It was fun to see him outside of Ohio, on an adventure far from our families. It felt like we were building grownup memories, beyond the proms and football games. Disney World is a place for kids and adults to be kids, but EPCOT felt more sophisticated, with pockets of the world to explore in glamorized sparkly Disney magic. As I held Ryan's hand, my feet yearned to go to the real-world places they represented. I knew there was more than these movie-set versions.

We ended up at the Morocco pavilion and explored the small bazaar as Casey took ten to talk to her boyfriend on the phone. She planted herself on a bench and waved us away, putting a hand over her free ear to block unwanted noise and catch every word.

The bazaar had gorgeous small rugs, lanterns, and clothing befitting Princess Jasmine, or young girls who wanted to be her.

"I love this place. All the colors and textures," I said over my shoulder to Ryan, whose arms were folded around me from behind.

"We should pick out something then, go on," he gave me a little nudge toward the stalls. I sauntered through the goods and flinched at every price tag I turned over. Everything was so beautiful, but I had only saved enough money for a small trinket.

"What about something like this?" Ryan drew my attention to a wall covered in brass lanterns. "That's kinda clunky to carry in my bag all day," I replied, scanning the assortment. But a mobile caught my eye and my ears as it tinkled in the hot Florida breeze. It had a brass dome at the top, and from there, six elephants marched down and around, their feet adorned with colorful beads and small bells. I picked it up and played with it. If I tucked the elephants up into the top, it would fold up easily enough to fit in my bag.

"It is really pretty," I admitted before reaching to put it back on the hook. It was forty dollars, more than I could spend, but I didn't want to tell him that.

"Why don't you get it?" Ryan asked, stopping my hand before I could let go. "C'mon, I'll get it for you," he said, starting to tug it from my fingers, but I pulled it in toward my chest.

"We'll split it," I said, and we did. Our first purchase together besides movie tickets or a meal. It struck me as strange that it was a decor purchase, something stable and steady for a home. I folded the elephants into my bag and we headed out.

The sun shifted down in the sky and the humid breeze carried a small chill. I held his face in my hands for a final kiss. I watched him until he was out of my sight and enjoyed the weight of the elephants in my bag. I could hear their bells tinkling with every step.

Evelyn swipes at the bells again, and the sweet jingle sings to us. I rock her and look at the calendar hanging next to the mobile. I mentally check off days. Only three weeks left before we head back to Hawaii and Ryan. He has been away for two grueling months of training at Marine Corps Embassy Security Group at Quantico, Virginia, to become a member of the Marine Security Guard. I'm glad we chose to spend this time with my parents, but my heart is tired of being apart.

We trudge up the stairs to the kitchen and I put Evelyn in her high chair. I kiss her head, breathing in her sweet scent. At the end of this training, we will learn where in the world we will grow next. Marine Security Guard detachments protect more than two hundred US embassies, consulates, and diplomatic missions around the world. We could go anywhere. Every time I think about it I have a mix of terror and excitement.

"Hey girl," Mom greets me from the couch. She gets up to kiss Evelyn's cheeks and hugs me. We raise our coffee cups in solidarity to start the day.

"Do you still feel like going to the fabric store?" she asks.

"Yeah. I think I finally settled on an idea," I say. I have been looking at the elephant mobile across from my cot for over a month. Evelyn loves it so much that I want to translate it into a quilt for her. I tell Mom I want to do rows of small squares and then bigger blocks with cut-out elephants stitched to the top. I want it to look like elephants are marching across it.

"Are you still planning on packing your sewing machine in air goods then?" Mom asks, pressing her coffee cup to her chin. For an overseas move, we will get an unaccompanied baggage shipment—our air goods. The rest of our items sail to us on a boat, which can take months. Our air goods, six hundred precious pounds of familiar belongings, will fly to us immediately, the seeds of the home we will plant in a new place.

"I am. I keep thinking I don't need to add that weight, like maybe I should pack more of Evelyn's toys instead, but ... "

"But you need something for yourself. It's okay." Mom grabs my hand and we lock eyes. They are misty. She is brave for me like I am learning to be brave for Evelyn. I want to know how to be brave for myself too.

I want to ensure Evelyn is happy anywhere we go in the world. But this means I need to work at my own happiness, and creating things makes me happy. The quilt can serve as a project that will keep my hands, head, and heart busy in the midst of waiting for the rest of our stuff to arrive.

As we eat our breakfast and Evelyn bangs her spoon on her tray, Mom and I sketch squares and jot down measurements. My heart slows from anxious pounding to a steady drum while we work. I can create beauty in this time of uncertainty. The elephants will march.

Back in the basement the next day, I spread the fabric by color out across the floor. The palette of light pink, mint green, and powder blue will be tied together with a soft gray fabric printed with silver palm trees. Touching the tops of the palms, I close my eyes and remember the chill of the hospital room where Evelyn was born.

It was raining that day. I held my new baby in my arms as drops drizzled down the panes, and underneath the window Ryan slept curled on the couch. As I watched both of their sleeping faces, something new grew in my heart. A root once meant for a family of two burst into three.

When the rain tapered off that day, I watched sunlight spread across the room, warming all of us. We became a family in that Hawaiian hospital, knit together with impossible love. I believed then that we three could be home anywhere together.

Now as the last rays of the sun cascade across the panels of fabric on the floor, my phone vibrates. It's Ryan.

"Hey, hon." I can almost hear him grin. "Morocco."

"Wait, what?" My heart thuds. "That's where we are going?"

"Yup. Rabat. It's the capital. I wanted you to know, but I gotta go. Love ya."

I barely squeak, "Love you too" before the line is silent.

I sit back on my heels and immediately pull up Google on my phone and type "Rabat." I hit the images tab and get lost in pictures of immense sandy beaches, souks, lanterns, rugs, women and men in long colorful robes. It is the real-world inspiration for the Morocco pavilion at Disney World, and we will become part of the daily life there.

In the morning, Mom and I get to work. She is helping me precut all the squares and will ship them to me.

"I'll wait for your address, and then we'll send it over," Mom says. Her hands are already busy folding and stuffing the pieces like a puzzle into the flat-rate shipping box.

"Okay. Thanks." My body feels numb today. Our unanswered questions about our family's next step have been abruptly solved, but a whole slew of new questions have erupted in my brain.

Where will we live? Will I be able to work? What is healthcare like? What about friends and community? Is it safe?

The list grows in my mind but I stuff it down, smashing it into compartments like the fabric Mom stuffed in the box. This is my fourth military move with Ryan and yet it is still daunting. I expect myself to handle it like an expert, but every time we shift locations we are at a different season in our life. This will be the first time we move with a child.

"Mom? I'm scared," my voice squeezes out, small and needy. I press my shoulder into hers as we stand side by side over the package.

"I know. But you can do this." She squeezes my wrist. "Remember that saying? 'How do you eat an elephant?' "

I put my hand on top of hers and answer, "One bite at a time."

Checklist

RYAN, EVELYN, AND I ARRIVE at the front gate of our assigned Moroccan housing at eleven o'clock at night. The street lights illuminate large swaths of vines that run up, down, and all across the walls of the property. Evelyn is sleeping, heavy in my arms as we swing through the wood and metal gate. A small tiled path and two marble steps welcome us to the front door. We struggle with the array of locks, requiring multiple keys, but eventually we are in.

The front hall features a giant crystal chandelier. My legs and back ache as we climb the winding marble staircase guarded by black metal rails. I take note of the provided furniture, but mainly my mental beacon is blinking for a crib. It rests on the third level on a bright red carpet in one of the three bedrooms.

The crib is wooden with a gray finish. It looks safe and snug. I can hear Ryan behind me, fumbling with our suitcases in the upper hall. He is digging for the sound machine. By the time he finds it, my arms feel like they are breaking, but I keep swinging Evelyn from side to side in front of the crib.

"I got it," he whispers and bends to plug it into the wall. "Shoot."

"What?" I say.

"It's the two-prong plug. Not American style. I forgot."

Our sponsor, Jeremiah, peeks in the door. The embassy has assigned us a sponsor upon arrival to act as our guide, helping us acclimate and get settled in Morocco. Jeremiah is tall with black hair and two small children of his own. His presence as we scramble in our new environment is both comforting and strange. He drops his voice to a whisper.

"Oh! We have adapters for you guys. Hold on."

His shoes plunk down the marble staircase and back up. He exchanges a plug with Ryan. The sound machine blares to life, and Evelyn stirs only slightly in my arms. Ryan and Jeremiah back out so I have room to put her down. I lay her into the unfamiliar crib and smooth her damp, sweaty hair.

I do not bother trying to change her clothing. As I bend down to kiss her, she lacks the sweet baby smell I know so well. Instead, she is a stale cocktail of sterile airports, plane seats, and body odor. Thousands of miles and snack crumbs cover her. They mix to make a musky odor, a badge of courage, a victory won: We made it here.

When I back out of the room, Jeremiah and Ryan are waiting for me in the hallway. I start to close the door but realize we don't have the monitor plugged in yet. I keep the door cracked and turn to examine the winding stairs. They look more like a death trap than a luxury.

"She won't be able to climb out of the crib, right?" I ask Ryan.

Jeremiah smiles. "We installed baby gates before you arrived. Let me show you." He swings a wooden panel across the top of the stairs and bolts it securely. This barely quiets my anxiety.

"Now, for your bedroom." He motions up yet another small set of stairs and we follow.

"We put all the sheets on your bed already and washed your towels." He also explains that he and his wife put some necessary breakfast items and two premade meals in our fridge and stocked our bathroom with toilet paper. Ryan and I thank Jeremiah again and again, leaning into each other, holding one another up with polite exhaustion.

"You guys look beat. I'll leave you to it." He and Ryan exchange a pick-up time in the morning for work. We walk back down the stairway to see him out.

We watch as the door swings closed, and I fiddle with all the locks. I pause to press my hand against the door. Ryan rubs my shoulder, pulling me back and into him.

"We made it," he says.

He tilts my chin up and kisses my lips. "Let's get some sleep, babe." He turns and gathers the rest of the bags and heads up the stairs. Before following him up the stairs I find the switch for the glistening chandelier. I shut it off and everything goes dark.

It's three in the morning when Evelyn's terrified screams wake me. Ryan has to go to work today, so I rush in to settle her before she wakes him too. Her little body is shaky in her begrimed airport clothes, and her eyes are wide with confusion and uncertainty. I pull her out, check her diaper, and proceed to shush and sing while I wiggle her into a pair of warm, fuzzy pajamas. When she starts to drift to sleep again, I place her in her crib. I pull a pillow and scant blanket up next to the crib and lie down on the floor. I fall asleep singing our own version of "Lavender blue, dilly dilly, Lavender blue. I do love you, dilly dilly, I do love you."

"Mama! Mama?"

I wake up with my hips aching from the hard floor and my spine crushed against the bars of Evelyn's crib. My eyes are dry and blurry. Evelyn's little hands push at the tops of my shoulders. She reaches to pat any part of my exhausted form on the floor. I roll over and hold her small, chubby fingers.

"Morning, baby," I say.

Evelyn giggles and pokes her pajama feet at the top of my head. I crawl up from the floor and gather her in my arms. I hear Ryan clanking downstairs. He is already dressed and toying with the coffee machine. We are lucky that our housing came with a welcome kit, three giant foot lockers full of basic household items.

Jeremiah went the extra mile and laid everything out for us

in the kitchen. I remember he said he and his wife went shopping for us. I pull open the fridge and find a dozen eggs, butter, yogurt, juice, and milk. There is a large Tupperware container with spaghetti and meat sauce. Garlic bread and a regular loaf of sliced wheat lie on the counter.

"Do you want me to make some eggs?" I ask as Evelyn reaches out for Ryan. He cuddles her cheek and kisses mine.

"No, there's no time. I'll figure it out there." He hands her back and goes to fight with the front door locks and alarm.

"Hang in there, guys," he says as he turns to leave.

The silence is deafening in the house after he goes. I envy that he already has a purpose today, a place to go and people to meet. I smooth Evelyn's hair back, and her bright blue eyes gaze happily into mine. So it begins.

We trudge back upstairs to dig in our large green suitcase to find her pink sippy cup and Minnie Mouse plate. A tiny fork and spoon come out next, wrapped in a bib. I place it all in the pieces of her highchair. I feel a little like Mary Poppins, pulling out all these things from our suitcase. It was awkward to pack them back in Hawaii, but now, on this first day, I am so thankful for my foresight.

Back downstairs, I put the chair together and attach it to our borrowed wooden dining seats. I go through the paces of working to make a regular morning for us, despite how cold and foreign the house feels. Evelyn plays at my feet in the kitchen as I attempt to use the unfamiliar stove. We are rewarded with a plate of spongy burnt eggs and toast.

A long marble hallway connects the kitchen to the dining room. The hall has two marble steps, so I carry Evelyn to her chair first and safely buckle her in, fastening the bib around her neck. I walk back for our plates, and one more time for milk in my coffee.

We eat in content silence. I look around the living room. Bright red carpets stretch across the cold marble. Plump but small yellow sofas provide seating. There are several hideous gold lamps shaped like pineapples. A giant china cabinet anchors one wall. Huge cream curtains are pulled to the sides of the windows, secured by golden hooks topped by cobra heads. The windows let in a bounty of light but are covered securely in iron bars.

Evelyn's high chair gives her the safety I need to run back upstairs alone and grab my journal and my phone. Back at the table, I punch in the code and remember: there's no Wi-Fi yet. I don't have any overseas data either. A landline phone is plugged into the living room wall, but I have no idea how to dial the embassy. I set my phone down and pick up a pen. It is barely seven.

I write down a quick list to motivate myself:

Explore House
Read Books
Explore the Backyard
Get Showers
Make Lunch
Evelyn's Nap
Unpack
Warm up Spaghetti Dinner

Leaving the journal spread open as a guide on the table, I wipe the crumbs from Evelyn's face and we dive into our tasks.

I tick off all the boxes leading up to Evelyn's naptime. When she is tucked in, I finally flop down on a yellow sofa. Her monitor whirs beside me. It was a fight keeping her awake this late in the day with her jet lag, and yet also a fight to get her overtired body to slump in my arms and then to the crib.

My own body shudders into sleep but I'm woken by a strange beeping sound, almost a song. I shoot up from the couch and run to peer out the peephole in the front door. There's no one there. A huge vine-covered barrier wall separates us from the street.

I tiptoe up the hard gray marble stairs to look into the video monitor connected to the front gate. There is an older man there, perched in front of the screen with a paperboy cap pulled low against his sunglasses. We were warned last night not to open the door to anyone we don't know. If someone is coming by, the embassy will tell us and the person will have a proper badge. There's no way I'm letting this guy in.

The door alarm jingles again. I grip the video screen and watch the man. He shuffles his feet impatiently. He pushes the call button and rings again. The sound is amplified in the empty stone hallways and walls. I hear Evelyn shuffle on her monitor and my heart clenches. He lifts his fingers again, and the chime swallows us. She lets out a tiny wail.

I grab the phone that connects to the outside intercom.

"Hello?" He looks directly at me through the camera and speaks wildly with his hands. But it is a tumble of Arabic. I can't understand anything.

"Sorry," I say and put the phone back. He pushes the button again, and though I know he can't see me, I duck down under the video screen. Evelyn lets out two long and loud cries. I curse under my breath. He knows I'm here. Will he leave? The bell rings again, and I sit up to watch him. His fingers are pressed to the button. The piercing sound does not shut off. It loops around and around, shattering all hope for quiet.

At the front door, I fumble with the locks. I fling myself out into the garden and toward the front gate. This man needs to leave. I slide the bolt on the gate and wedge my foot in between it and him. I have no words, so instead, I lift my hands in what I hope is an international sign for "What is the issue here?"

"I am the landlord. You have a problem with your pipes. I fix." He pushes against the door.

"No. There is no problem with the pipes." I push back.

"Yes, yes, yes," he says and pushes harder.

"No sir!" In an instant, my right hand is on his chest and I push him back onto the sidewalk. I slam the door with my left, bolt it, and run up the stairs. His angry shouts follow me.

I put the front door between us and race back to the camera to watch him leave.

Idiot! My brain shouts at me. *What the hell?* Every movie I have ever seen has taught me to know better than to go to the door. I know better than to open it. *Why did I do that?* Evelyn's monitor is on the floor. She's quiet. I grab it and slump back onto the couch.

The living room phone rings. I rush over to it.

"Hello?"

"Lindsay?"

"Ryan?!"

"Yes, it's me. What are you doing? I heard our landlord was at the house." Apparently, word gets around fast at the embassy. I tell him about the misadventure.

"You pushed him?" he asks. I can hear his disappointment. Day one and I'm already making waves for him.

"Well, I—" Hot tears well in my eyes. *You fool.*

"Listen, it's fine. The staff is smoothing it over," he says and explains that the embassy personnel are apologetic. They gave the landlord the wrong house number. Someone's pipes are broken, but not ours.

As I hang up, Evelyn wails again. I walk to my worthless cell phone on the table. She has given me a scant forty-five-minute nap. There are two things left on our to-do list for the day. Onward and upward. I force my legs back up the spiral staircase.

It feels so good to sling another empty suitcase out into the hallway. My clothes and Ryan's clothes settle into the closets and drawers. My toiletry bag sits next to the pink sink in the bathroom. Looking in the mirror, I see the skin below my eyes is still puffy with jet lag. I smile back at the tired mom anyway and give her a mental high five for pushing through this day.

"C'mon, Evelyn," I say and scoop her and her suitcase into her bedroom. "Let's see what you have in here!"

We unzip the bag and she squeals at seeing her tutus and fairy wings, superhero capes, her favorite fuzzy puppy dog, and the baby doll that has never earned a name other than "Big Baby." She squeezes each item up to her small face. I'm thankful for deeming such items as life rafts of happiness for our first few months here.

I'm also thankful these items will keep her busy for a minute. I pull out each tiny top, bottom, dress, and jacket and refold them in tidy stacks in her new closet. I bend down to grab the final pile out of the suitcase and look up to see her disappear behind the bathroom door across the hallway. As I stand up, the massive wooden door swings shut with a distinct click.

"Baby!" I run across the hall, skidding in my sock feet to the door handle. I pull it, and it's locked.

"Babe! Evelyn!" I take a breath. "Evelyn, can you come turn the lock for Mama? The shiny button in there?" I realize my two-year-old is not going to figure out what I'm blabbering about. But I see her shadow near the door.

"Mama?!" She tugs at the handle. "Mama!" She begins to wail.

I steady my voice to express a calm I don't feel. "Evie? Sit down for Mama. Sit on the floor. I'm going to get you out."

I search the nearby rooms for keys. There aren't any that fit the lock. I run to my phone frantically and curse myself for the second time today for no data, no plan, and why didn't I ask Ryan how to call the embassy back when he called me earlier? Shit. Shit. Shit.

Evelyn's screams echo off the walls. I had not even checked out that bathroom yet. What if there's a chemical in there or soap to eat or something else hideous? I kneel next to the door and examine the lock. It's a flat line that looks like it could fit a knife.

I'm down and back up the spiral staircase in seconds with a butter knife in my hands. "Hold on, Evelyn, hold on," I say, crouching down on my knees. I stick the blade into the flat bolt. I jiggle it and turn it to the right. The lock makes another click. I pull at the handle and fall through the door on my hands and knees. Evelyn is curled up in the dusty bathtub, tears streaming down her face, the fuzzy puppy clutched between her hands.

"Mama's got you. I got you," I say, pulling her out of the tub and into my arms. "We're okay."

I manage to finish unpacking the suitcases and leave them in a deflated pile downstairs. My head aches with fatigue but it's time to sort out dinner. For the first time, I realize we do not have a microwave. I pull out a skillet and pan from the welcome kit and examine the stove again. The burners are gas, easy to light, and I warm up the sauce and meat. But the oven? It is gas as well, but when I turn the knob, no pilot light flickers on. I can smell the gas, but obviously, I need to light it myself. There's a red-handled lighter in the welcome kit.

"Evelyn, babe, you gotta get back," I say and hover on my knees by the oven door with the lighter. I gently tug the trigger on the lighter. Out from the oven floor erupts a burst of flame. I smell burnt hair and touch my left eyebrow. Perfect. We'll have piping hot garlic bread, and I just received an eyebrow singe for free.

Bing-bong! The front door alarm goes off and I turn to see Ryan in the doorway. I crumple back on my heels and smile. I can't wait to tell him about today's adventures with landlords, locks, and lighting ovens. He catches me and Evelyn up in his arms.

I relax and relinquish the day. I surrender to this moment where we are back together, building toward the feeling of home.

Connections

WHEN RYAN DROPS A WHITE SMARTPHONE in my lap at dinner, it practically glistens with new possibilities. Our house does not have internet yet, but this phone—this is a beacon to the outside world! With my new phone, I have access to *data*, and the word itself has never tasted so sweet on my tongue. There have been good lessons from this month of being jolted away from the internet, but tomorrow morning I'm setting my pioneer existence aside. I'm plugging back in.

I hunch over the phone for an hour, entering passwords and establishing email. I log in to all my social media accounts. I delight in each notification "ding!" Like Gollum from *The Lord of the Rings*, I have found my "Precious" once again. Once I soak up the latest and greatest in all of my feeds and forums, I'm left with jittery fingertips on the keys.

With one device, I reconnect to the world around me. I'm able to video chat with my Mom and Dad at the click of a button. I schedule a time to virtually see my best friend and sister later in the day. Once Evelyn is up, we snuggle on the couch and watch YouTube videos together. Why must Cookie Monster always think the letter of the day is a cookie? Hilarious!

My music account is back and our favorite tunes follow us through the house. I look up a new recipe for dinner and we walk to the local market and get supplies. This is a banner day. But while Evelyn naps, I find I am still scrolling, only with less excitement.

I'm reconnected to the world around me. But I'm not connected with anyone here. While talking to my family and friends from far away has helped, it is not the same as in-person friendships and interaction.

Ding! A new notification arrives, and my finger swipes eagerly. I've been invited to a mom group in Rabat. I squint down at the person inviting me. I recognize her name from my husband's work roster. I am *in*. I've discovered the door that all my potential friends have been hiding behind. I accept the invitation and swipe.

Hungrily, I push up on the messages and fill myself in on the group's chatter. They are scheduling playdates. They are sharing funny mom memes and inspiring parenting articles. These people could be my people. But then I see that there are no playdates scheduled for this week. I bite my lip and swallow back the nagging mosquito called insecurity. If I listen to her, she'll suck all the bravado and fun out of what must be done.

I type: "Hey everyone—we just moved here. I'm a mama to one two-year-old girl. We need friends. We have limited toys right now but we have boxes! We have bowls and spoons! Is anyone up for a playdate this coming Friday at ten?"

With one tap of my finger, the message goes out onto the waves of the group chat, my floating hope to make friends. I can masquerade as a mom seeking a playdate for Evelyn, but really it is a playdate for me. I am desperate for actual human contact. I can see when people read my message, and I imagine them wavering at the decision to come to my house. A total stranger who just moved here. Maybe I am too eager. But then one mom chimes in: "We'll be late, but we'll be there!"

After her, another mom responds and then another! Pretty soon I have a real playdate party on my hands. I'm excited, I'm overwhelmed, and it's time to run back out to the mini-mart and pick up—what? Cheese? Crackers? Whatever! I will make this work.

Two days later our door buzzes and the first few mothers trickle in with their adorable children. We circle up on the rug, bouncing babies and filling our plates with snacks. The toddlers take bites from our plates and crumbs quickly cover the red carpet. We exchange our names, our kiddos' names, and where we all came from.

The conversation turns toward Evelyn and me. One of the mothers asks if I will be placing Evelyn in the *creche*, which is the French preschool program.

"Oh, I don't really know," I say. "I don't think so."

Ryan and I have already ruled out the all-day French school-slash-day-care. We only have eighteen months here in Morocco and a lot of family and friends are coming to visit. Putting Evelyn in all-day care does not feel right for our family, but I'm afraid to share too much.

"Well, are you hiring help at least?" one mom asks.

"Um, I had not really thought about it. Probably not."

"Well, you won't be able to do much around here without school or a nanny," she says. The conversation turns back to the group. I am out of the hot seat, but I have a few burns.

When the moms and babies clear out, I am left with crumbs and questions. I don't know what I am going to do here or who I'm meant to be now. The playdate left me with a few potential friends and the realization that I am not in my comfortable norm anymore. In Hawaii, most of my friends stayed home with their littles. It was a fairly large community to pull from. I always had somewhere to go and someone to be with.

Ryan is the last doorbell ding of the night. He walks in smiling while Evelyn and I are at the dinner table, finishing up supper.

"How was the playdate?"

"Pretty good. Some nice moms." I sigh, and I see him take in my slumped shoulders.

"But?"

"Well—I'm just not sure where I fit yet."

"It'll take time. It was only the first meeting—and this might cheer you up." He pulls a flat-rate box from his bag.

I carry the precious box into the kitchen and trace my fingertips over my mom's delicate handwriting. I cut through the tape and carefully peel back the white flaps. Inside are the precut elephant quilt squares. I spend the rest of the night laying out the pieces on the rug. The elephants march together in a beautiful order. I sigh. I hope I find order here too.

The next week, Evelyn is balanced on my lap and Mom's face is on my phone, which is propped against a water bottle as I push the fabric of Evelyn's new quilt through the sewing machine.

"It's really coming together, Linds," Mom says, the video call audio crackling a bit. "How's everything else? Any more playdates?"

"Yes. Jen—the British mom? I saw her on Monday." I tell Mom about how we had a lovely afternoon at Jen's home, and that Evelyn and her little boy had played well with each other. We were able to sit and visit with one another and drink a coffee! But when she drove me back to our house, our toddlers were fussy. We started singing to them and ended up belting out "Part of Your World" from *The Little Mermaid* all the way up the busy street.

The song had calmed both the kids and us down, and as we sang the final notes with dramatic flair, we laughed. Being with Jen felt easy and I felt seen. I'm starting to understand those are two of the most comforting pieces of the finding new friends puzzle. When another heart recognizes mine and I can see theirs, even though we haven't spent much time together, it reminds me of *Anne of Green Gables* and how Anne discovered, "Kindred spirits are not so scarce as I used to think. It's splendid to find out there are so many of them in the world."

"We had a good moment. I think she could be my first friend here!" I say to Mom.

"Well, you just need one friend. Remember that," she says. "Everyone else will be a bonus."

I smile as I squint at the needle. Each row of elephants needs to be as straight as possible. Evelyn wiggles so I use my left hand to bring her in closer.

"I know. Thanks, Mom." Evelyn shifts again, almost sliding off my lap. "This is getting tricky, I'm going to go," I start to hang up but Mom puts her hands up.

"Let's try something. Evelyn? Do you want a book from Nana?" Evelyn nods and smiles, her whole body alive with excitement.

I set her down on the couch and prop Mom up next to her on a pillow. Mom holds up a picture book and starts the story.

I curve the seam down along the blanket and pause by the elephant's back leg to backstitch. That should hold it down. I do not feel like I have a firm grip on Morocco yet. I think back to our time in Seoul and how it became enchanting. My curiosity and delight there felt constant. Not here. Not yet. Morocco still feels like it is buried under layers. I am hopeful it will captivate me in the same way. Perhaps I need to choose better eyes to see it with.

A warm little hand presses into my arm. Evelyn is at my side and I glance at the couch to see Mom is still propped up on the pillow. From thousands of miles away, I still secured a few precious minutes of grandparent childcare.

"Thanks, Mom," I say, picking her up again.

"Sure! Show me what you've got done."

She tells me it looks great and I tell her how impatient I am to finish. Our household goods are still not here, and I have yet to understand who and what I will be here. With everything else out of order, completing this one project feels like a step in the right direction. It is filling the hours before new friends come to join our table. My hands and mind are processing how to pull a new life together, not just for me, but for Evelyn too. This little bit of stitching is pulling me through our rollercoaster. I ride up and down on the thread.

In the following week, I complete the quilt. Evelyn sits on my lap again as I finish the final stitches.

"Here we go, Evelyn!" I say and place her back on the living room rug. I spread my hand across the fabric to check for any extra pins, and then I fluff the quilt up into the air. It splays across the carpet. Her smile shines up at me.

"Mine?" she says.

"All yours," I say. "Check it out!"

She dive-rolls onto it and rubs her cheek on the soft fabric. I giggle and dive down to roll with her.

I am still in this body, this heart, and this mind. But I am changing once again to fit our new surroundings. I can feel an old me and a new me emerging in this house. We are tangled together, deciding what pieces to prune and drop so that new sprouts can emerge. Lying on my side, I trace the elephants with my finger. Finishing this quilt is one mountain scaled. I am ready for the next.

Vacation Pie

"THIS HERE? THIS IS VACATION PIE. Once you take a bite of this, I need you to be here with me."

Ryan holds out his fork, and my lips waver over the flaky crust and berry filling. *Can I do this?*

The loudspeaker in the café crackles. A voice announces something in both Arabic and French. I suck in a breath. In thirty minutes, Ryan and I will board a train and ride from Rabat to Marrakech; from there we'll travel ten hours to the Sahara Desert. I planned the itinerary. We will climb dunes, ride camels, and eat meals without Evelyn in tow. I did not plan to feel sick about leaving her behind.

With Ryan's support over the past few months, I've been finding a way back to myself. It started with having a babysitter come to our home once a week. While she plays with Evelyn, I move upstairs to our office, plop myself down in a desk chair, and write. With each word my pen unfurls onto the page, I discover fragments of my past and present that are beginning to fit together.

Some of what I have been finding hard to face is becoming clearer. I took up the role of mother with diligence, responsibility, and great love, growing better and more comfortable with it each day of Evelyn's life. But I haven't reckoned with our back-to-back deployments.

Between the deployments and training, our move to Morocco, and Ryan's current work schedule, I've let our relationship seep through the cracks of my fingers, ever so slowly, like a leaky faucet.

Drip, drip, dropping away. I imagined rifts in marriage to look like screaming, slammed doors, and miles of past hurts. But our quiet chasm has been formed by challenging life changes. We have yet to bridge the silence.

The piece of pie wavers between us. I remember how Ryan and I promised ourselves early on in marriage to travel. In the past, that became a centering place where we remember who we are—together. Evelyn is being well cared for back home by her grandparents. Under the table, Ryan nudges the toe of my boot with his own. The invitation hangs in the air. The Sahara is the one adventure Ryan has requested since moving to Morocco, and he wants to be on it with me.

He smiles, extends the fork, and raises his eyebrows. I take a bite.

The train seats are sticky with sweat. We ride past goat herds, small humble towns, and a few bigger, modern buildings. I alternate between reading a book and staring out the window. We sway back and forth in the heat of the car and observe the flow of local families boarding and departing as we pull into each station. Ryan reaches for my hand while he closes his eyes, and I put down my book to hold his hand.

It is night when we reach Marrakech, but it is far from dark. We taxi to Jemaa el-Fna, the main square in the ancient section of the city called the medina. A baggage man from the hotel meets us, and we have to jog to keep up with him through the crowds. I wish we had a second to pause in the thrum. The walls rise up higher around us as we go deeper into the heart of the medina. In a packed corridor, a moped screeches inches past my body. I have to suck my stomach in so the seat doesn't graze my skin.

"Did you see that?" Ryan asks when we reach our room. He closes the door on the noise of the night.

"Which part?" I say.

"That moped. It came so close to us," he says.

"I know. It reminded me of—"

"Thailand." We say it together. We grin.

At the hotel, I pull out our toothbrushes from my backpack. I load them up and hand Ryan his. Our eyes meet in the mirror as we brush our teeth together. It's a less glamorous routine we have kept up for years, and tonight, in this strange city, it feels comforting. As we lie down, he puts his arms around me.

"It's okay to miss her," he says. I nod into the pillow. I do miss her. But I have also missed us. I pull his arms tighter around me.

~

After a two-day trek by car we arrive in the desert. When I touch my scalp, there is sandy grit under my hair. As we step out of the jeep, Ryan pulls one of the two blue scarves from our bag. He insisted we pick them up from a roadside shop.

"A *shemagh* is essential in the desert," he says. He shares with me how they had them in Iraq and demonstrates how to tie it.

"Will you do it for me?" I ask, and he does.

The camp manager comes to tell us that the sun will set in an hour. Now is the perfect time to go up the dunes and watch the sunset. I follow Ryan's lead, but I am like a newborn fawn, my legs tripping and sinking in the sand.

"This is rough," I grunt.

"Tell me about it. You got this, though, c'mon," he says. We walk across the tops of the dunes, then straight up the highest one, where we perch and take in the endless brown wave tops. The sun starts to dip, kissing each dune as she goes. The heat follows her. I shiver and lean my body into Ryan.

"Just wait," he says. He keeps his eyes trained on the sun. She touches the horizon and bows gracefully out, but when she does, a whoosh of wind flies across the earth and blows through my hair.

"Whoa," is all I can say.

"Yeah, I remembered this. I was hoping you would get to experience it," Ryan says.

"It's like the world just exhaled," I say.

The corners of his eyes crinkle as a smile climbs up his face.

"Did I ever tell you about what the Egyptians believed about the sun?" he asks.

"No. Tell me," I say and rest my head on his shoulder.

"They believed the sun sailed through the sky on a barge from dawn to dusk and then rode down into the underworld at night."

"So the sun dies every day?" I ask.

"I guess you could see it like that, yeah," he says.

Ryan insists we sleep in hammocks under the stars that night. He shares stories I have never heard about his deployments to Iraq. Before I drift off to sleep I think of the Egyptians and the sun dying each day. Some pieces of me have come alive but others have perished since having a child. In tending to her needs, I have forgotten mine—ours.

When we get home I will kiss boo-boos, read books, cut up peanut butter and jelly sandwiches, and dole out snuggles. But someday that won't be enough, and someday she will leave us. I need there to be an us when she leaves.

The next day we journey to an oasis in the desert. Camels gallop toward us when they see us coming. They wait for our guide to pump water into a trough for them. We're told each camel belongs to a nomad family, but they roam free on their own until they are needed. We watch them nudge one another out of the way for drinks, pulling their heads up and sloshing the water around their mouths with satisfied grunts and slurps.

It is too hot to even want to touch one another, so Ryan and I spread out on individual mats. He closes his eyes to rest, and I lie down on my belly with my pen, ready to pour my thoughts into my journal. Now that we are alone together, I wish we had tried harder to make time for us a priority. I wish I had tried harder. Ryan's jaw goes slack, and I look at his sweet face. I love his face. His chest rises and falls in sleep, and my chest tightens.

How could I forget about him? Forget about us?

I have needed to do the work on myself, and yet I don't want to do this work and lose him along the way. I know the military will continue to throw us curveballs. There is so much we cannot control or choose. I know we can be far away and miss each other, but the opposite is also true. We can be together and miss each other all over again. We have a choice in that. I close my journal, roll over on my mat, and submit to the desert wind and camel grunts, a distorted lullaby.

When the train rumbles away from the Marrakech station, the sand has been scrubbed from our scalps and skin. We're in an air-conditioned car this time and have had a refreshing night's rest in a comfortable hotel room. Our days in the Sahara were unforgettable and exhausting; it is hard and good to think about going home to Rabat. I can't wait to hug Evelyn, but I don't want to go back to how we've been living. Ryan gazes out the window and I pull my journal from my bag.

"Are you ready?" I ask.

"No," he says, and then turns his head back to me with a small smile. "But okay."

We decided to use the ride back to talk through some questions for couples I found last night on the internet.

"What is your love language?" I ask.

"You know that," he says, and I nod.

"I wanted to make sure it was still—"

He puts up his hands. "It is. It's acts of service." I look back down at my page and the words blur. This feels staged. Forced. Pointless. I close the book and look at him.

"I don't know where I went wrong, but I know this hasn't been good," I say, gesturing between us. He doesn't answer at first. In response I turn my body to look out the window, but I look without seeing. Instead, I turn to watch our moments in the desert, replaying newly formed memories in my mind.

I had reached for his hand on the train ride down here, and we had clinked our glasses together over our meals. I want to remember how I looked back at him from my camel, his face relaxed and content as he rocked from side to side on our ride through the sand. He even tried sand surfing down a dune and effortlessly cruised from top to bottom. I cheered for him all the way. On the last morning, we woke up for the sunrise. He pulled me from bed so we could witness it together. I sang "Happy Birthday" to him because he turned thirty, and I got to be the one there to celebrate with him.

Ryan interrupts my gaze out the window, drawing me back toward him with his voice. "It's not just you," he says. "This is an us thing." I turn toward him, my body and my heart. We talk about how it is easy to love our daughter and how wonderful that is. But our love, we're learning, takes a different kind of work.

We have to continue to pour in. To make space. To hit the refresh button. To forgive and rebuild, flirt and battle, cry and celebrate. We have to care.

"I'm not giving up," I say.

"Me neither."

We grab and squeeze hands. It feels like we are beginning again. We will begin again as many times as we have to, to keep going.

Inshallah

THE ESPRESSO MACHINE PUFFS and steams as another customer places their order. With our babysitter tending to Evelyn, I am an island to myself, on my own for the day. Snippets of conversation swirl around my head, but I cannot understand any of them. I can gather some of the bits in French, but the Arabic hits my ear like the trombone voice of Charlie Brown's teacher: "Wah, wah, wah. Wah wah? Wah wah! Waaaah."

I cannot help but turn my head to the conversation next to me. Even after almost a year of living in this country, moments like these shout out my foreignness, reminding me I am not fully assimilated. Still the foreigner without foreign language skills. The men chatting are a contrast. One is fully clad in a stark white traditional *djellabah*. The long robe-style tunic touches the floor. His friend wears a sharp navy blue European business suit.

The mat in front of me is covered in crumbs from the *pain au chocolat*. My cappuccino has the perfect amount of froth and a welcome touch of cinnamon sprinkled across the top. This little coffee shop has become my regular haunt. It is adorably French, down to the rustic flooring and the case of delicate pastries. The waiters wear jaunty white hats and blue neckerchiefs.

An hour ago, I pressed my back into the cream-colored couch at a counselor's home office. I booked the appointment because this past month I have felt like I'm melting, Wicked-Witch-of-the-West style. This has happened before; my worries accumulate, roiling beneath the surface of my facade. I need to deal with these emotions before I dissolve.

In the past, I've been hesitant to seek counseling. I felt it was weak to need professional help. Instead, I leaned on friends, family members, working out, and burying stress far beneath the surface. But lately, I can't surmount this feeling that pulls on the edges of my nerves. Each day has been a test of resiliency that I greet either as a withered green blob on the floor or as a screaming banshee.

The counselor is also a military spouse, which is one of the main reasons I reached out to her. I think that our shared experience as spouses will help when I explain my downward spiral of emotions. And I hope that this will make it easier to be open with her. She knows the landscape of this life.

We broke into the session with me making light of everything. When I was growing up, my parents always had it together. Their inner turmoil and worries stayed far away from my sister and me. I am ashamed that I cannot do the same. Then again, my parents never moved to foreign countries, raised a kid overseas, or faced immersion in military culture. I have to stop comparing lives.

"I want to stop feeling," I say, finally getting to the truth. "I'm exhausted by my emotions."

The counselor runs her fingers through her dark hair and meets my eyes, where tears are already threatening.

"You are carrying a lot in your family's current journey. There is not much certainty or control. We are going to try an exercise in which I am going to challenge you to see your anxiety as a friend."

Anxiety is my friend. A friend? My mind twirls for a minute before sinking back into puddle mode. I want to wallow. It is more comfortable to lie in my green pool of unhappiness. It is easier to do nothing.

Ryan warned me this appointment might be difficult. He had held my hand last night and reminded me we are a team, but that

I need to hunt out different tools to handle my stress. He cannot keep leaving the house to do his job and worry about me. I can do it for him. I can do this for my girl. I nod at the counselor to continue.

"Close your eyes and place your hands comfortably in your lap. Take a long breath. Now let's visit all that is pulling you downward."

I crack my eyes open to give her a squinty glare.

"I know it is painful," she says, "but we are going to walk with your feelings today instead of suppressing them."

I squeeze my eyes shut again.

"Your emotions are drowning you. Let's grab onto them as a life raft instead and float back up to the surface. Greet each new problem. Do you see one? Grab it! Pull it in and let it lift you up. Are you at the surface yet? No? Okay, keep pulling them in as a little raft around you."

Damn therapy. Why do I need to be the one who solves this?

Inhaling and exhaling deeply, I surrender to the darkness. My eyelids tremble with tears as I greet each one of my weights and wish them into buoys.

We might not get to go home for Christmas because they have not issued our plane tickets yet. Why not? I want to go home. I need a break. I want to be somewhere that just feels normal. Where will we go after this? What country is next? When will Mom and Dad get here? Will they be able to come if things in the world stay so uncertain? Oh, I miss them. My sister Bethany already bought her ticket to meet me in Ohio. She's trying, and I'm going to fail her.

I am failing all of them. What happens if we have to evacuate the country? Ryan will have to stay behind. I'll have to take Evelyn with me, alone. I took the peanut butter out of our go bag the other day. I need to add more snacks, and does she still fit the clothing I put in there four months ago? If we don't go home for Christmas I need to box up everyone's presents and send them out. I hope our family's gifts get here on time. I am glad I pushed to put up the tree anyway.

I pull the random thoughts around me on the couch and use them to build an imaginary life raft. I am riding them up to the surface.

"Take a final breath in, and out, and sit with all of your worries, concerns, and emotions," she says. "They are a part of you, and it is all right to take time to feel them. Do not push them down. Exhale the breath. Open your eyes."

My cheeks are hot and wet. I shiver and meet her eyes. She reaches her hand across the space and grabs mine.

The heavy weight that was pulsating on my chest is less crushing. I envision my green puddle draining into the cracks of a stone floor. I rise mentally and emotionally on quivering, but stronger, legs. For the first time, I've been guided through my feelings in a way that felt nourishing. What would I be like if I continued doing this kind of work? I don't have answers to my worries and doubts, but I am paying attention to what is bothering me instead of pushing it aside.

As we close, the counselor reminds me it's going to take time and work to learn how to process and move through my anxieties. But as I leave our time together, I gauge how I'm feeling. I feel relief and empowerment to take another step forward. I know it will be worth the work.

Another blast of steam erupts from the espresso counter, calling me back into my journal pages. I scribble down a few last lines and down the remaining cappuccino from my cup. The men give me brief nods as I head out the door and into the bright sun.

On my way to the car, the tall, tan minaret roars with the midday prayer. It's poor timing for my departure. Cars swerve and park haphazardly on the sides of the road. Men and women leap from their vehicles. Everything stops, everything is interrupted to remember Allah. Stacks of cardboard stand outside storefronts. Men take breaks from their jobs and drag the cardboard out into the middle of the sidewalk. They find Mecca. They crouch down, knees pressed into the makeshift mats, and close their eyes. It amazes me that this happens five times a day. It amazes me how public it is.

Ashhadu an la ilaha illa Allah.

I bear witness that there is no god except the One God.

God is in my life privately. He's with me in a corner of the house each morning, my head buried in a devotional. He is with us at the dinner table when we thank Him for the food we eat. I throw prayers up to Him as our day goes along. Our journey with Him is quiet. I have never wanted to be noticed for my faith. *Do not be the Christian who barks out to God and tries to convert others.*

But Morocco shows me the noise. The daily scramble for faith. The devotion to kneel openly in the street and pray, to pull over from driving and enter a mosque. By law, the minarets must be the tallest buildings in every Muslim village, town, and city. The call to prayer reaches across the land with nothing to stop its resonance. There is no way not to notice it.

It's chilly so I settle into the driver's seat and crank the heat. I will wait for the roads to clear before shifting into gear. Meanwhile, the men rush out of the restaurant. I watch the white robe and the navy blue suit disappear inside the mosque. As I watch them, I wonder: *If I had more faith, would I handle my anxieties better? Is counseling the answer? Is prayer the answer? Can it be a mixture of both?* I tuck my chin into my chest and send my own silent prayer up to the heavens for grace to handle whatever comes next.

~

Allahu Akbar, Allahu Akbar, Allahu Akbar, Allahu Akbar.

I wake up in our cold bedroom and pull the blanket up around my chin. December in Morocco has no Christmas snow but it is fraught with frequent rainstorms. The house is built for the blazing hot summers. All cold marble flooring, cement walls, and large, poorly sealed windows with bars on them to keep the cold in. The frigid air soaks into my bones. It reaches up through the floor and turns my body to ice.

My eyes feel like cotton balls, dry and puffy from crying. The call to prayer persists, bouncing off our bedroom walls even before the first shard of light enters the room. It's official. We are not going home for Christmas.

God is Great. God is Great. God is Great. God is Great.

When Ryan came home late last night, I already knew something was wrong. The front security gate creaked open at his touch. The keys rattled outside as he hunted for all the locks on our giant wood and metal-clad front door. I knew I should go to him and greet him. But I stayed with my face pressed against the yellow rental sofa, gazing aimlessly at some random show on Netflix.

I had called a few times but he never answered. I thought about ringing the main post at the embassy, but that would be a poor reflection of who I am supposed to be. As a military spouse, our superpower is to be the one who understands. The mission comes first. Always—duty first.

I turned to see him in the doorway. His clothing was rumpled from the long day. His face was weary. As he told me the news, I cried and clenched my fists, but I didn't cover Ryan in anger. I didn't blame him like I might have in the past.

This thought wakes me. Throwing covers aside, I slide on a sweater and my slippers and step gently down the marble staircase, tiptoeing as quietly as I can into the living room. The Christmas tree glitters from the corner, welcoming and warm. My body feels stiff and tired as I creak down into the overstuffed chair.

I pull a blanket over my legs and allow my journal to fall open. I have been submitting my writing for a few months now to various online publications and need to prioritize some opportunities that have come through. Yesterday I wrote out my to-do list and I added: Finish up work deadlines *(just in case)*.

I am good at covering the uncomfortable unknown in tasks. *Just in case.* I was prepared for the worst. And it came. We are all still here. While I was scratching out my small tasks yesterday, big decisions were being made. The world feels too close right now, rubbing into me uncomfortably. We learn that the US Embassy in Israel is moving from Tel Aviv to Jerusalem.

Because we live in a Muslim country, we have been encouraged to tread lightly and limit all unnecessary travel. Our trip is canceled. World politics surrounds us like a hotbed of coals. There is no sheer will or determination that will get us on the plane back to the United States. No brazen thought like, *This is my right.* It is hard to explain this to our family and friends, nestled in their stateside houses. I am jealous. I imagine some people in the US might hear this news and have the luxury to turn it off when the negativity and unrest become too much. I imagine I have done the exact same thing before—but now there's no turning it off for us.

I click my black gel pen and get to work. As much of a home and community as we have built in Morocco, we are still guests in a foreign culture. The quiet of the house is soothing. I center myself with the techniques I am practicing from counseling. Sit with the pain. Let it wash over you. Don't try to blink past it too fast.

I start to make new lists of places in my life where I can be grateful. This loss of a trip home does not need to be where our light is snuffed out. The pen touches the page and I let the words flow:

> *I wanted to see my sister, and she's already bought her plane ticket. I wanted the familiarity of Mom and Dad's cozy house and giant cuddly Murphy dog at my feet. I wanted to hug my nieces, nephews, and cousins. I wanted to see my grandparents and for my daughter to experience Nana's magical Christmas home decor, her magical Christmas Eve*

with the giant white and blue tree and the warmth of her meal and after-dinner coffee. It would have been nice to relive that magic with my own family. I wanted to sit together and watch the snow drift down upon the cornfields. I wanted familiar shops, food, and FAMILY surrounding us.

But that's not what happened. It's not okay and I do not have to make it okay. However, we must move forward with what we DO have.

Closing the journal, I feel a thin strand of resiliency start to stretch around me. It is easier to be fearful and angry. I can stay stuck, silently cursing this holiday season and our circumstances. Or I can teach myself something about hope. I can remember it is not "us" and "them." This is simply the way it is.

Evelyn cries from upstairs. The neighborhood donkey cart clip-clops along the road beside our house as I climb to retrieve her. She is standing up in her crib, brown wisps of hair sticking up all over. She smiles and reaches for me.

"Hi, baby girl." I pull her up and into my arms and hold her small body tightly against my own. I soak up her warmth. She always radiates what feels like living joy.

"Are you ready for breakfast?"

"Scambled eggs. Toast. Pwease, Mommy." She grabs my face with her chubby toddler hands and plants a wet kiss on my cheek.

I fulfill her order and the rest of our morning is a slow one. While I have resolved to make this time a good one, that does not happen with a Mary Poppins spoonful of sugar and a snap of my fingers. Exhaustion blurs the edges of my movements. Roller coaster emotions can be as tiring as running a marathon, except there's no endorphin high at the end. Evelyn and I work together to unpack and put away the clothing I had packed for the trip home.

As I place each item back on a hanger or on a shelf, I let go a little more. Finally, we shove the travel shampoos, medicines, and first-aid bag back into the master bathroom drawers. Evelyn plays on the floor next to me with a box of Q-tips, spreading them across the floor like snow. Peering out the window I can see our car down below, parked under the peeling white and green awning. Then I remember groceries.

Ryan had left me the car. He knew keeping busy would be the balm to get me through the disappointment of our holiday plans.

We get dressed; I make an extra effort to wear something bright and happy, smoothing on my rebel red lipstick and pulling my hair into a low bun. Evelyn picks out bright pink pants and a sparkly blue long-sleeve shirt. I pull her hair into pigtails and we giggle together as we brush our teeth, our matching blue eyes meeting in the mirror.

She claps her hands when I give her a kiss, transferring a touch of lipstick from my lips to hers. We will not shrink today. We will walk into that grocery store with jingle bells in our hearts.

I yank a cart from the stack and we're off, pushing through the giant Marjane doors. We lovingly call Marjane our "Moroccan Walmart." Everything you need for a one-stop shop and plenty of what you do not need. The store is huge, with rows upon rows of grocery items, housing goods, and clothing. I realize I have not been here since the end of November. I stumble to a halt when I see a giant *Joyeux Noel* display. Forty-four years of French occupation in Morocco have given us the unexpected gift of Christmas.

"Kissmas tees, Mommy! Look! Look!"

"I see them, bug. They are beautiful. Let's see if they have a star, hey?"

Our tree at home has yet to have a star placed on top. How perfect would it be to find one here today? Excitedly, I push the cart through a giant festive display of jam. Ferrero Rocher chocolates are piled like a block of gold up to my waist. I see something in the back—a wall glitters with Christmas ornaments and strings of lights. We get closer and Evelyn sees them first.

"Staaahhhs, Mommy! Staaahhhs!" There are three colors: red, silver, and gold.

"Which one? You pick, baby. We'll put it on the top of the tree with Daddy tonight!" Her pudgy finger pokes the gold one.

"Good choice!" Placing the star right next to her in the front basket, I am high on the fumes of finding Christmas in this store. I squeeze her small body. We dance down the aisles and collect the rest of our items. I sing "Jingle Bells" and "Santa Claus is Comin' to Town" softly to her as she claps her hands. I select a giant container of croissants. Why not? I let Evelyn pick out a small box of intricately decorated Moroccan cookies from the bakery stand.

We have great fun examining the *Joyeux Noel* seasonal meat case. It has interesting items in it: hooves, tongues, livers, hearts, and even gray, rippled brains. I place chicken and beef into our cart. The local women shopping around us give us big smiles and wave at Evelyn. She ducks her head into my chest.

In the fruit and vegetable area, I grab brown paper bags and start to fill them up one by one: potatoes, bananas, carrots, cucumbers, eggplant, cilantro, and apples, shaking the dirt off them as I go. I push up to the weigh station. Each sack needs to be weighed by the store employees. A little white sticker tag scrolls out from their machines with the weight and cost.

I start to leave the area but a woman in a bright pink *djellabah* gently presses her hand into my right bicep. She motions to the women working at the juice stand, stationed in the middle of the fruit section. They are waving at me and beckoning us over. The head juicer says something in Arabic first. I shake my head.

"Je suis désolé. Je ne comprends pas," I say, apologizing for my lack of Arabic.

"Ahhh," she says and tries again in French. *"Jus d'orange?"*

"Oh, non, merci." We do not need orange juice today.

Her face wrinkles in concern. She motions to Evelyn. *"Non? Pour l'enfant? Jus d'orange?"*

I realize this kind woman just wants to offer Evelyn orange juice.

"Oh! Oui! Merci! Merci beaucoup."

I am not sure if I am supposed to pay for it or not yet until I watch her. She whirls around the small rectangular stand, pulling out a whole orange, rinsing it, and pressing it into her machine. Her hands shake a bit as she pulls back, looking for a travel cup to pour it into for us. She settles on an old raspberry container and tries to pour the liquid into the flimsy plastic square.

"Mmmm, delicious!" I say, "Here, baby, take a sip."

Evelyn thinks this whole exchange is simply wonderful and claps her hands. She leans forward and tries to drink from one of the floppy edges. A little dribbles off her chin onto her dress. I'm not sure how I am supposed to carry this mess out of the store, so I smile back at the juicer.

"Excusez moi?" I hand her the rectangle and make a show of pulling Evelyn's sippy cup of water from my bag. She nods eagerly in understanding and quickly sets the sloshing juice rectangle on the countertop. I unscrew the sippy cup lid. She takes it from me to pour the water out and the juice back in.

Handing it back to me, our eyes meet. I wish I could tell her what this small exchange has meant to me today. I will my eyes to pour appreciation into hers. Evelyn laughs as she receives her cup back, full of juice now instead of boring water. I whisper in her ear, and she looks up at the woman.

In her tiny voice, she says, "*Merci*."

I look up at the woman's happy gaze and utter the Arabic word of thanks I am pretty good at, *"Shukran."*

The juicer nods at us and finishes our exchange with *"Inshallah,"* a common response to just about anything. It means "God willing."

After a day of letting the pain of what we do not have knock around in my brain, I am starting to remember what we do have. Evelyn's eyes gleam up at me from the cart, which is loaded down with a festive bounty, including our precious tree-topping star.

~

Allahu Akbar, Allahu Akbar, Allahu Akbar, Allahu Akbar.

God is great. Another morning call to prayer finishes up as I squeeze some drops into my dry eyes and shuffle down the stairs. This morning, Ryan got up with Evelyn and they spent some quiet time together while I slept a little longer. I find them both wearing superhero capes and greet them, wrapping both into a huge hug. They hold out a mask and cape for me.

"Come get me, Mommy!" Evelyn cries, tossing her shimmering purple cape over her shoulder.

I pull the black felt mask down over my eyes and race after her.

"Boo! Gotcha!" We tumble down onto the deep red carpet in the living room, a tangled heap of limbs.

"C'mon, guys, these look good." Ryan carries out the cookies we purchased at the grocery, along with a steaming pot of mint tea. We pause to sink our teeth into the delicate flavors of almond, orange blossom, and tightly wrapped, honey-soaked baklava.

I reach across the table and place my hand on Ryan's rough cheek. I brush a small crumb from the corner of his mouth and he gives me a questioning, hopeful smile.

"You better?" he says, placing his palm over my hand.

All that comes to mind is *Inshallah*. God willing. God willing, I will be.

From Glimmer to Shine

I STAND AT THE EDGE of the green astroturf with Laila and Pragna as the students filter into the play and education center. The green beneath my feet reminds me of the way I connected with this opportunity in the first place. It was a serendipitous meeting in a backyard garden at a children's birthday party. There, with a lush emerald lawn beneath my bare feet, I was leading Evelyn through different activities set out for the kids.

The birthday party was for the son of the counselor I had met with, and the party was at her home. Sometimes overseas life tends to blur the edges between friendship, professional workspaces, and brightly colored party balloons and cake. I could either be unnerved to have all of these interactions bump together, or I could give it a shrug of acceptance. I smiled at friends at the party while realizing I had also sobbed on the cream-colored couch in the corner of the living room during my counseling session.

I remember my counselor/friend came up to Evelyn and me as we were playing.

"I want to introduce you to my friend Laila when she gets here," she said. "She's creative, like you."

"Sounds fun," I said, smiling.

When Laila arrived and introductions were out of the way, I learned she ran a children's music company called Crescendo Baby Music. Her classes taught young children the joys of early exposure to music education. She was interested in opening an experience for older children as well. She explained that she wanted to put together a musical theater program for kids in Morocco. As we talked, we played with our little girls, teetering between motherhood and professional interest. As I listened to her new dream, I explained that I had experience teaching dance to young children and that I used to dance professionally. I told her I missed it.

She asked what I thought of working with children with very little technique or background in dance. I explained what I knew to be true, that the children I worked with gained confidence and a love of the arts as they moved through my classes. A select few may go on to participate as professionals in the dance or theater world, but that's not the primary objective. Helping create patrons of the arts and individuals who continue to support and enjoy theater is an important impact as well.

Her smile spread and we exchanged numbers to talk more.

Later, after a lengthy discussion over the phone, I sat on the floor while Evelyn played kitchen. I looked over my notepad, covered with ideas and underlined letters in all caps. I remember I sat back on my heels and watched Evelyn scooping dry beans and colorful pom-poms out of a bin with a little spoon. Was this my next right step, teaching these classes? If so, who would watch Evelyn?

Laila had proposed that the commitment would include meetings with the other Crescendo teachers, as well as smaller group sessions with her and her colleague Pragna. My portion of the program would require me to teach one out of every three weeks.

Scoop, scoop, pour. The beans in Evelyn's bin tinkled down into a small orange plastic soup pot. She stirred them carefully, mixing them with the pom-poms.

The way Laila saw it, each teacher had a specific set of skills we could offer the musical theater program. She would be the music director, Pragna would oversee acting and writing the script, and I would be the dance director.

"Look what I made, Mama!" Evelyn said, bringing me her soup pot.

"Mmm, delicious," I said after taking an imaginary bite. I pulled her onto my lap and kissed her cheek. "Thank you, honey. What else can you make me?"

Evelyn bounced back up to start making again. I marveled at her, the ease with which creative play flowed from her. I wondered what I would be like if I allowed myself to feed my love of movement along with my writing here in Morocco.

When I discussed it with Ryan in our kitchen, he didn't understand why I was questioning it at all.

"You should say yes," he said. We were scrubbing and drying our dinner dishes. I washed a soup bowl and handed it to him.

"But what about Evelyn?" I asked.

He rinsed the bowl carefully before drying it with his towel. "What about her?" he asked. "We'll figure this out."

I bit my lip. It seemed easy to say things would work out when the mechanics of figuring out timing, people, and places would be my responsibility.

"But we don't have a full-time babysitter, we don't have—"

He dried his hands and placed them on my shoulders. I took a breath as he explained that I could drop off Evelyn at the embassy. She could play in his office for the short while that I would be teaching, and then I could swing by and pick both of them up on my way home from class.

"I can do that?"

"We can do that."

I dried my hands and watched him leave the kitchen. The program would span the final months of our time in Morocco. It would culminate in a final performance just a week before we would pack out. And I could lean on Ryan's help.

As I hung up the towel, I realized I didn't trust that we would work it out. With the military's fluctuating needs and schedules, I haven't been able to rely on Ryan for childcare. It isn't his fault, it just is.

Or it just was. What if I trust that we will make a way?

I look down at the green astroturf between my feet and breathe deeply. It has been about two years since I've taught a live class. I hung up my dancing shoes in Hawaii for the deployment and Evelyn's infancy. But as I follow Laila and Pragna into the class circle to introduce ourselves and the kids tell us their names, I'm ready to give it all I've got.

I warm them up, helping them prepare their bodies to move across the floor. While there are one or two kids with some training, I can tell most are new to shifting their bodies into these stretches and shapes.

I don't want them to lose confidence or give way to self-consciousness. I move us into a series of acting exercises where they use their bodies to demonstrate happiness, sadness, boredom, and anger. I have to use my whole body too, for while the class is learning English, most of the children speak *darija*, colloquial Moroccan Arabic, as their first language and French as their secondary. I am humbled that their lives incorporate three languages, where mine requires only one, even while living overseas.

The French I studied for three years is a bonus—and sometimes comes in handy, but I am far from fluent.

We communicate mostly with hand gestures and eye contact, and I pour energy into every action. Our session ends with instructing a short few eight-counts of choreography, more acting-based than dance-based. They rise to the challenge, and as the music from Disney's *Moana* fills the space, they pretend to raise their oars in imaginary boats and sail through our first session with strength and grace.

A smile lingers on my lips for days, as does the ache in my body! It feels wonderful to be back at it, to share this joy once again. Ryan notices it too. My attitude at home is smoother and more balanced. Having this little piece of dance and performance in my life—sharing it with others and watching them grow—rekindles an ember I thought had long gone out in my heart.

Over the months in the classroom, I've wondered why I let go of teaching in Hawaii, but what I needed in that season was different. I had no idea this gift would come along, but I'm glad it has. Not every post has a specific placement for my skill set or can work with our family priorities. What is available and what I'm good at are sometimes two entirely different things. This can be a frustration, or it can be an opportunity to grow in a new way.

In Morocco I've learned that saying yes has opened a whole new door for me. I'm interacting and invested on a deeper level. I was worried about leaving Evelyn behind while I teach, but she is well cared for every time I leave. At home, she watches me create dances and joins in to offer her ideas. I am glad she gets to know me in a new light, a little glimmer of who I was before her, glistening through to who I am now.

Evelyn also has to join me at team meetings sometimes. One day we are in Laila's lovely apartment. Laila, who is Moroccan, welcomes us into her bright living area. Her long, curly hair bounces behind her. There is a team of musicians scattered across the couches and floor to practice the music for our end-of-program showcase. Most of these musicians are local, plus one who is French, and they work with Laila in her main programming for Crescendo Baby Music. I marvel at how each one is skilled in multiple instruments and ideas.

While we run through the planning of the show, Evelyn has already taken off to a back bedroom to play with Laila's girls. Soon Pragna arrives with her son, and he's off to play as well. I stand to welcome Pragna, and we do the customary kiss-kiss on both cheeks, followed by a hug.

Pragna is from Canada, with warm brown eyes and smooth, thick black hair. She pulls the recent script edits from her bag and we sit together on a couch to run through scene changes and a new creative rap she's written for the kids. The students have been doing an excellent job of learning their lines. The show is in English, as the parents hoped for them to practice this skill set. I admire each time a child stands to work through a piece. Their tongues tangle on different sounds, but they smile and keep trying. Pragna is an effervescent cheerleader, smiling and consoling as they work.

One of Laila's girls flies from the back hallway, shaking a tambourine, with a parade of children behind her, all grins and giggles. Laila hugs her, and the musicians strike up a song for all of us. The kids dance and stamp their feet. Pragna and I start to clap, getting up to dance as well.

These are gifts of teaming up with Laila and Pragna—bearing witness to their art and how they bring their love and skills to their family. Our kids share in the process. Each of the four kids at the meeting has mini tantrums and snack needs. We take turns caring for them, in community together. One of us moves them back to build blocks. Another brings out a snack tray. We let them disassemble the couch cushions to make a fort. Several of the musicians offer them opportunities to play gently on their instruments.

I have always thought of work and family as two separate entities, meant to be held separately. But I'm witnessing a new way. A way that acknowledges we are all layered in our different talents and the ways we love. And why not? Why not let the kids into the room? Why not let Evelyn behold that I am both her mother and—dare I own it?—an artist who needs to create.

A few weeks before our performance, Laila arranges a field trip for the class to attend an orchestra concert in downtown Rabat.

She wants the children to see inside a professional theater, practice sitting in the audience, and hear a full orchestra play together.

I help chaperone, which mainly means sitting behind the kids and alternating between smiling back and forth with them, shushing them, and signaling for them to sit on their bottoms. I let the music lift me and take me back to all the productions I've had the pleasure of experiencing, on the stage or in the audience.

"Art is a reflection of life," a teacher once told me. "When we cannot find the common words to express ourselves, we act. When acting isn't enough, we sing. When singing cannot reach the depths of emotion, we dance."

During the grand finale, the seven kids in my section burst from their chairs to their feet. I start to pull them back down, but they are moved by the music. Deep percussion thunders through our bodies. Lilting violins hum. The children's eyes are wide, their bodies tilting forward on their tiptoes, soaking in each bite of sound.

Their souls gulp at the nourishment, art, and beauty surrounding us. One child lifts his arms, matching the movements of the conductor. The others notice and lift theirs too. Their limbs sway with power, and I can't stop myself from both crying and laughing. The kids are overcome with life, moved to stand and feel part of the art before them.

It's the best dance I've seen in a long time.

On the day of the show, I stand backstage with Pragna and the kids. They are in their places, ready to begin when the curtain opens. We can hear the audience on the other side, the great hum of a crowd settling in, waiting and wanting. I know Evelyn and Ryan are out there too, and I'm sweaty with my own nerves, ready for them to see what I have invested in. This is what our creative scheduling and time blocks at home were working toward.

Relentless energy radiates through the kids. They poke one another and whisper, and Pragna and I motion wildly at them to quiet their voices, still their bodies. We try deep breathing techniques and shoulder pats and show them how to ground their feet. We point at the microphones that carry their murmurs out into the audience. The excitement is overtaking them, mixing with anxiety. This is their moment, sink or swim.

Laila gives the kids a final pep talk before stepping out to the front of the stage. As she speaks into the microphone, a wave of silence finally falls over the kids. When her speech is finished, the audience puts their hands together for the first time, and I turn to watch the kids' faces. Even in the dark behind the curtain, they glow with surprise at hearing applause that is just for them. They beam, toothy grins and wide eyes, and I smile watching them.

I know it will go by too quickly, the lights and scenes and songs whisking them up into a world we built, and then they will bow and it will be over. But the feelings—this moment before the curtain parts, the quiet cold of a hushed audience, the thunderous applause after a performance—will stick somewhere in their bones, a memory of anticipation and the gift of offering a practiced performance. I move to the wings and delight in watching the show from this vantage point, how the energy in their bodies carries them forward to say the lines, move into choreography, and sing songs.

The grand finale is a rendition of "When I Grow Up" from *Matilda the Musical*, and I hug myself listening to their voices lift. The lyrics swan dive into my heart:

And when I grow up
I will be smart enough to answer all
The questions that you need to know
The answers to, before you're grown up

It makes me laugh, a small chuckle in the base of my throat. I don't have the answers, and I am beginning to understand I never will. There is a great mystery to life, and part of it is leaning into what I can't know and holding tight to my truth.

Living here, experiencing this program and these kids, and embracing this opportunity has offered me another chance to step into the light. To share my own, and to bask in the shine of others. At the end of the performance, the teachers come out to wave and take a bow, and I see Evelyn sitting on Ryan's lap, clapping hard. I wasn't sure how we were going to work this out. I was almost too scared to try.

I smile at Ryan and he waves at me. We did it.

Laila comes to visit my home the week before we pack out. I've already said farewell to the rest of the team during an evening out on Le Dhow, a restaurant on a boat, docked and gently rocking at the mouth of the Bouregreg River near where it empties into the Atlantic Ocean.

That night we sang "You've Got a Friend in Me" over the crowd and danced on the deck as the sun dipped low beneath the sky. I hugged everyone and had the painful twinge of knowing this might be the last time I saw them in person. I got in my car that night and said, "Thank you." I whispered it into the steering wheel, into the starry night sky, into my heart as I edged away from this chapter of my life.

Now Laila sits on my couch, and I thank her for this time with her team. She thanks me as well, and lets me know if ever I come to Morocco again, she's here. I tell her I know this, and we grab hands.

"Please, please, let your light shine. Don't dim it, my friend," she says, and I nod. I know I'm a better spouse, mother, and human when I follow this advice. This adventure has proven that I can step away for a time and our family doesn't break.

When I wave goodbye to her from my doorway, I press my hands deep and hard against my chest. I wish I had another year, another round of classes, another performance to cheer on here. But I'm leaving with my light on, and for that I am grateful.

The Worry Box

PACKING TAPE EMITS an unmistakable sound as it rolls through a dispenser, screeching along to seal the top of a cardboard box then making a harsh *thunk* as the tape is torn from the roll with a quick pull. Our home echoes with these sounds, along with the crinkle of brown paper and the scuffle of the movers' feet.

My sandals slap up and down the marble staircase. One hand brushes the wooden railings while the other traces along the cold, white cement wall. I inhale deeply through my nose to calm my heart, working to steady myself as I watch our items being jammed into plain square boxes. As layers of our worldly treasures are folded in, my thoughts cling to them as I try to forget about them.

I feel the first shudder of starting over as our personal items leave this home. I know this feeling well. Where I once could catch a whiff of my husband's aftershave in the bedroom, it is now rife with the pungent odor of cardboard boxes and sweaty movers. I slide out of the way as another man walks into the room with a stack of boxes. I think I smile at him, but it might have been more like a grimace. There is a tense surrender in having movers here to pack us.

This is a job we could do for ourselves if we were stateside. But I relinquish control and watch the boxes roll through the door and up the path from our Moroccan home to a large moving truck. From there it will go to a port. And I have to trust that somehow it will sail across the world and reach us in Quito, Ecuador.

I make my way down the hall into our daughter's room. On the bed, I drop a bright red piece of construction paper on which I had scrawled an explicit "NON" in my chicken-scratch handwriting. I grin at the polite afterthought added to the bottom: "*Merci*."

I never did get any better at Arabic. My make-do French still reigns.

The sign is to tell the movers not to pack what is beneath it: Evelyn's warm bundle of blankets and a pillow with her beloved Bunny tucked haphazardly underneath. Bunny's gray stitched eyes peer out. Even she seems to know the next airplane ride is near.

Fifteen months of my little one's life is in this room, her hands pulling at the drapes to reveal morning sunshine, her dreams filling the dark. I pick up Bunny and rub her well-loved ears. I hold her to my face and catch the sleepy lavender scent of my girl. I trace the plush nose, remembering the many times this past month that Evelyn has woken upset, both she and Bunny wet with tears.

The crying began after a nap. I remember that morning I had been sorting toys and clothing while Evelyn played beside me. For every toy I tried to put away, she found another to fill the empty space I made. I managed to laugh at the difficulty of sorting and prepping for a move with a robust three-year-old. I found I could make headway in fifteen- to thirty-minute sprints before needing to switch my attention back to her. We played outside that morning, racing and rolling in the grass. Lunch was nothing special. She fell asleep peacefully but awoke crying so desperately I thought she had hurt herself. I pounded up the stairs, opening the door to find her curled into a ball, clutching Bunny to her chest.

"What is going on, honey, are you hurting?" I asked, coming to stroke her forehead. She shook her head. As I pulled her up toward me, sticky snot streaked across my shoulder. We sat down in the oversized yellow chair in her room. "Was it a bad dream?" I prompted.

She looked into my eyes and held Bunny closer like she was trying to attach her as a third limb. "We move today, Mama?"

I pulled her closer. "No, not yet sweetheart." We stayed together for a few minutes, her body relaxing as an onslaught of thoughts slammed into my brain:

Could my three-year-old be concerned with the upcoming move? Obviously. What had she heard Ryan and me saying about it? Had I been positive in my words? Negative? Aloof and unaware of anyone's feelings but my own?

She rubbed Bunny's soft ear across her cheek as I kissed the top of her head. I wanted to ease the pain of all this change.

Should we have been hiding the move from her? I made a deal with myself that later I could grab my phone and search internet articles about moving, child psychology, and best practices for not messing up our kid.

I carried her downstairs and we tucked into an episode of *Daniel Tiger's Neighborhood* with a side of cinnamon graham crackers. As Daniel zipped around our favorite neighborhood on his trolley, her tears dried. I relaxed and began to think that maybe this would be a one-time occurrence. There was no reason to panic yet.

But it happened again.

And again.

And again.

Naptime or morning time for the next few weeks meant Bunny, tears, and her repeated question, "Are we moving today, Mama?" Sorrow in the heart of my small, sweet human. When I shared her pain with family and friends, they told me there was no way she could be clued into the large PCS challenge in front of our family.

And yet I began to see she was clued into me.

When Evelyn was twenty months old, I had sat with Kaylynn at our dining room table in Hawaii. We talked in hushed tones about deployment rotations, her upcoming move, and how, while we dreaded change, it was always the one constant in military life. We reached across the table to hold one another's hands as we started to cry. Evelyn came to my side.

"Mommy, buddy?" she said, holding up a stuffed dog for me to hold.

"Oh, thank you, baby," I said, wiping my eyes.

She studied my face and said, "Mommy, candy?"

I burst into fresh tears, which fell in surprised joy. But with this new bout of crying Evelyn came closer.

"Mommy, coffee?"

Kaylynn let out a window-rattling laugh. "She knows you!"

I carry that moment with me.

I am delighted that my child could, and can, comfort me, but I am confused too. *Am I allowed to show my weakness? How much crying in front of my kid is too much? When is it damaging? When do I hide it? Do I need to hide it?* I can't remember my mom crying. Or my dad. Or my grandparents. That doesn't mean they were right, or I am right, or that either of us is wrong. So much of parenting is navigating answers we don't have in the moment. We only know what we know, and we do our best with that knowledge. But I long for guidance on what is good. Good practice for our hearts, for our minds, for our souls as we navigate challenges together as a family.

Military life asks so much of those who serve—and of their families. We are now responsible for helping our child process and persevere in this life. *Is there an untapped resource, some hack I could be doing to get us through the hardship?*

Can I be a mother, a military spouse, and myself, tangled between bold and breaking? Can I love the adventures of this military lifestyle but also feel exhausted with each new trial?

I give Bunny one last snuggle and place her back on Evelyn's pile. As I make my way downstairs to watch more boxes roll out of the house, my phone bleats from my back pocket with notifications. I check it and grin at a photo that has just arrived: a smiling Evelyn is chopping vegetables in my friend Helen's welcoming kitchen, shielded from this moving chaos.

In the photo, I can see the happiness in Evelyn's eyes, the true bliss of being scooped up by someone who loves and knows her, and I am thankful that she's been released from the high-pitched squeal of packing tape and the rustle of brown paper. Thank goodness for friends who live alongside us in this transient life, who swoop in to make things easier when they can.

I met Helen through Jen a mere twelve months ago. I remember our feet sunk into the lush grass of Jen's backyard as we pushed our squabbling two-year-olds on the swing set. We recognized the glimmer of a real friendship from our first conversation. I survey my backyard from our upstairs office window. We had picnics with our kids there. There were birthday and Christmas celebrations, baby showers, simple days at home together, and adventures out where we toted the kids through familiar sights. We've laughed through hard days and kept each other calm when our children melted into tantrums.

Drawing closer to the window, I see our front gate. I also remember crossing our backyard path, taking steps toward understanding that I am and can be more than a mother. With their encouragement, I walked down that path to greet Helen and Jen for ladies' dinners out on the town. A new portion of Morocco opened for me after I met them, a sweeter portion, the one that comes when friendship blooms and nurtures the place where rest and growth happens. I am grateful for this chapter with them. I'm also already mourning a future without them. They are diplomats as well and will also leave Morocco in time. I don't know when or if we will see each other again.

When I miss a place, I miss the people most.

My phone dings again as Helen asks how the packing is going.

I tell her that we have about two hours left. She sends a thumbs up and a smiley face. When she brings Evelyn home, the house will be empty, aside from our suitcases and the embassy furniture. Our fifteen months here feel like the blink of an eye, the flicker of a hummingbird's wing. As we leave Morocco, I will look back on pictures of people and places and wonder if it really happened.

This was the place I believed would never feel like home, but somewhere in the middle, I changed. I realize every single place I have lived has looked me straight in the eye and demanded: *Will you be open here? Will you allow your heart to spill out and learn what you can from those you meet along the way? Can you accept that you will never be the same?*

Maybe this is what I get to teach Evelyn; maybe this is what I still need to learn.

Maybe it is okay to grieve, to let tears fall, to witness repair in relationships, to see the people I love move toward wholeness as I struggle for strength myself. The world is wide and wonderful, but I have to stay open to feel it. The good and the hard. I don't always get a choice in what each chapter delivers, but I have agency in how I choose to handle it.

The call to prayer sings from the nearby minaret. Maybe it is also okay for me to reach for something more. I make my way downstairs to the living room and find the side table, which has its own red "NON" sign. I slide open the drawer to pull out my journal and devotional.

The movers wave at me and somehow convey with gestures that they are loading the last of the boxes but will take turns to pray. I nod and motion that I'll sit here and wait for them to finish. I nestle myself into a chair and crack open my devotional. I watch as two of the movers take unused cardboard out to my backyard. They place it in the grass and kneel upon it. As they bow their heads to pray, I bow mine too.

~

When Evelyn wakes up the next day, she is not crying, but she holds me tight as I pull her onto my lap in the chair.

"How ya doing, love?" I ask.

"Mmm, I'm 'kay," she says and squeezes Bunny.

"That's good," I say. She sits up straighter on my lap.

"My kitchen is gone?" she asks.

"Yes. Your kitchen left with the moving truck yesterday. We'll see it again though, at our new house," I say. She burrows into my shoulder. Not knowing what else to do, I tell her again what is coming. "Soon, we will get on a plane and fly to see—"

"Mimi and Papaw!" she fills in.

"That's right! And then we go—"

"To our new home."

"Yes," I say and smile at her. But she is leaning back and twisting Bunny's ears. "Remember what we saw about Ecuador?" I shut my eyes and think about the video I showed her of our next station. "Llamas and alpacas, great big mountains, the Galápagos Islands!"

"There will be sea turtles," she says and I nod.

"There will also be you, and me, and Daddy. That is the most important," I say as she sighs. We are quiet for a while, and it feels like the emptiness of the house presses in on us.

"How is your heart?" I ask.

"It feels like worry," she says. I nod, my chest tightening.

She leans against me and I think back to my devotional yesterday and the cardboard boxes folded out for prayer in the backyard. I think about how faith was something I had quieted a long time ago, and yet somehow Morocco shook the need loose again. Perhaps hearing a call to prayer five times a day did it. But more likely it was because I had arrived in this country at the bottom of my reserves. All techniques I tried to quell my rising frustration and anxiety had faltered. When I started to look beyond myself, I found my soul starving for faith, and I had begun soaking it in slowly and intentionally through books and conversations with friends.

"Let's come sit on the floor for a minute," I tell Evelyn, and I guide her to a spot on the floor by her window. I pull back the drapes and sunshine spills across the carpet, warming the place where we sit. I place her across from me and hold her hands.

"All that worry inside you? It is inside me too," I tell her. "But we can't keep carrying it around." I think about how far I feel from harnessing unshakeable peace. *But is that what faith is? Arriving at a place where I would always believe things would pan out? Or is it more of an everyday wrestling with the truth that I can't walk this earth by my own power, and that I don't have to if I choose to believe in God?*

"Do you remember Jesus?" I ask Evelyn as she smiles and wiggles on the carpet.

"Yes, Mommy." We had been reading a children's storybook Bible each day.

"Let's make a worry box for him today." She lets go of my hands and cups a hand over her mouth.

"That's silly, Mommy," she says.

"You're right, but silly is good for us," I say. "Now, let's build our box." On the carpet, I trace a square with my finger. I ask her to help me build up the sides. She pretends to nail them in with an imaginary hammer. "Good! So next we're going to put all our worries in. Here, I'll go first." I pretend to write on my hand. "I'm worried about where we will live." I pretend to tear the worry from my hand like a piece of notebook paper and drop it into the box. "Now it's your turn," I say.

I know I do not want to use religion like a Band-Aid. But I do want to hold tight to what I believe could be an anchor, a place to rest. Maybe the great resource I have been looking for all along is the soul comfort of hope.

Evelyn places her hand on my palm. "I'm worried about the plane." I pretend to toss another paper in the box. We go back and forth like this several times, tossing our worries into the box.

"See all those worries in there?" I ask, and she nods. "Let's close up our box for now and lift it up. C'mon, let's get under it!" We pretend to close the box and lift it between us. "Up, up, up!" I say. I wave goodbye to it as if the ceiling has cracked open to the sky.

"Where did it go?" Evelyn asks.

"Up to God. He can carry it for a while," I say.

We walk downstairs to start breakfast. As I scramble the eggs, I think about how I want to have an answer for her, an easy way through these challenging times.

But the way forward is taking one step and then another. We won't have all the answers even after we arrive at the next station. We'll have to build the answers as we go.

I imagine our worry box floating up among the clouds. As I set our food on the table, we gather our hands and heads to pray. We can build faith along the way too.

Lost and Found

RYAN, EVELYN, AND I POUR OUT into baggage claim at the Boston airport like sludge from clogged gutters. The lights overhead are dim and the airport walls, floors, and chairs are a sea of muted gray. We shuffle along like zombies, our legs groaning from being shut up in the plane for so long.

"Babe, what time is it?" I croak. My throat catches on what feels like a pile of sand lodged in the back of it. My tongue is dusty too, and I'm certain my stale breath can take out an army.

"We have about two hours before our connection. Relax." His eyes are trained on the crowd; he is searching for the right baggage carousel. Now that we're back on US soil, we have to collect our baggage, clear customs, and recheck everything to continue on to our destination.

Evelyn zips ahead of us, her three-year-old legs thankful to be free. But she hasn't slept in twenty-odd hours and her energy is frantic. Her eyes are puffy and wild. I reach out to contain her.

"Just let her go. Stand here while I get the stuff." I appreciate Ryan's direct orders. It means I don't have to think. As a Marine, he is trained to function at high capacity on little sleep. As my husband, he knows when my edges are raw. He can focus on the mission, and I will gladly gather our carry-ons into a little pile beneath me and stay put. The baggage carousel sputters and luggage starts spewing out of its gaping mouth.

I look down and pick at a piece of crusty vomit on my jacket sleeve. There was turbulence coming into Boston and Evelyn's belly could not take one more change in altitude. The white paper bag had crinkled in my hand as I shoved it under her mouth just in time. Now, her pent-up energy crashes into the side of my leg. I try to bend and hug her, but she wiggles free and runs around me in circles.

The suitcases come around and I watch them with careful eyes. My back tenses into a knot. Ryan drags each bag over to us, and I begin to count them. He brings over numbers five and six. I shake my head to try and wake my fuzzy brain.

"Is that it?" he asks.

"That's all the bags, but where's the stroller?" I say, and an ominous signal flares in my brain.

Yesterday, the chandelier had twinkled over our heads one last time as we counted our bags in the front hallway. There were six suitcases, three carry-ons with wheels, two backpacks, and a stroller. The stroller was packed in its own clever travel bag with an easy-to-carry black strap.

I had surveyed everything on the cold marble and tried to find space for the rolled-up elephant quilt inside each of the bags, but all were full. After waiting so long for our household goods in Morocco, I've learned the value of packing well. I tried the backpacks next, but they couldn't hold the quilt either. I chose the best option and shoved Evelyn's sweet elephant quilt into the stroller bag.

At the check-in desk at the Rabat airport, the attendant was confused about our military orders. He waved his hands over all our bags.

"We're moving," Ryan explained.

The attendant simply shrugged and hauled them up for the weight. He tagged them and flung them onto a conveyor belt. I watched them disappear through the dark black flap.

"We can check that for free," he said, pointing to the stroller, which we had planned to carry on.

"Don't you want that?" Ryan asked me.

I paused. Evelyn is so big, and we already have so much to carry—to Paris, then Boston, then finally Columbus. It was going to be a rough thirty-hour slog. I made a quick decision.

"We can check it." I thanked the attendant, who tagged it and shoved it through the black flap. As we walked away, I remembered the quilt and gasped.

At the baggage carousel in Boston, we count the suitcases over and over and turn again to the carousel, still rotating but empty.

"Maybe there's an oversized baggage area," I say to Ryan.

"That's a good idea," he says and walks off. Evelyn finally collapses into the pile of luggage at my feet. He returns shaking his head.

"We have to get checked in for the next flight," he says and slings a backpack onto his shoulders. I nod, but the knot in my back has turned into a heavy stone.

As we drag ourselves to the next checkpoint I touch his arm.

"It's not the stroller I'm really sad about."

"Oh?"

"Her elephant quilt? I put it in the stroller bag."

His eyes meet mine. Tired heaviness weighs down the corners of his mouth. He doesn't say anything else, just squeezes my arm.

We fall into a bitter sleep on the next flight. Evelyn's head bobs against her chest. Ryan lets out quiet snores. My eyes have turned into the same sandpaper as my throat. We wake at our final stop: Columbus, Ohio.

We collect the bags again and haul them down the carpeted hallway. We are all exhausted, shaky with fatigue. It is nice and strange to see Ohio State memorabilia everywhere, but the noise around me is overwhelming. I hear English everywhere. After so many months of being insulated from background conversations in languages I didn't understand, my brain is on fire, trying to process the familiar words flying at me.

I hold out hope at the baggage claim. We count and recount the suitcases. The stroller is still not here. Out on the curb, we search for Ryan's parents. Their familiar gray van pulls up, and I hand Evelyn to her Grandmother Mimi. Evelyn gladly throws her arms around her neck. My stomach clenches as we drive away.

Morning comes, and I silently slide out of bed, trying not to disturb Ryan. I tiptoe quietly through the kitchen to the coffee pot. I look down and marvel at the cups lined up next to it on the counter. My father-in-law is an early bird like me, and he knows I like my first cup of coffee piping hot, accompanied by silence.

He has filled the mugs on the counter with steaming hot water so that my cup is already warm. This small gesture wraps me in an early morning hug. I inhale the aroma of the fresh grounds from the prepared coffee pot. I dump the water from the mug and watch it swirl into the silver sink basin.

It is homecoming day. When we come back to the United States, everyone wants a piece of us. It is tough to spend our entire time in the car driving for visits. It ends up feeling like coming home is work instead of a much-needed respite from overseas life.

We've learned from those past visits, and this year we have invited everyone for one big homecoming day. Food is spread across all the spare inches of the counter. I can tell my mother-in-law has been preparing for weeks. Another warm wave covers me. It is a lot of work to fly back and forth to see our people. It must be equally hard to wait for us to return and host us well while we are here.

I pour coffee and shuffle to the fridge for milk. I pull a spoon from the silverware drawer on my way there. Even after months away from this place, I still know where everything is. Milk retrieved, the next important step is to decide between sitting in the plush great room or heading out to the porch. I choose the porch as the weather is good, a balmy spring morning, with sweet birds singing. Their songs settle my busy mind. The wooden deck boards are rough against my bare feet, grounding me. The Adirondack chair is perfect against my sore back. Even though travel days in the airport are sedentary, my body carries the compression of altitude, heaving luggage, and change. I imagine myself like a tight coil, pressed down but lacking the exciting spring.

Twenty minutes go by, then the door swings open behind me. Ryan takes my cup inside. He comes back with his first mug and mine refilled. His face has a touch of stubble, and I am grateful for these few days when he will not have to shave and tie up combat boots.

With one hand on his coffee cup, he reaches across with the other hand. He interlocks his fingers with mine. I don't have to look at him to know what he's thinking. After all these years, the view of the property is still the same.

We drink in the view and our coffee. The tall pine off to the left of the deck sways in the spring wind. A chime dangles from one of its low branches and sends soothing tones into the air. The pond in front of us, dug by Ryan's great-grandfather Bud, is alive with cicadas and frogs. Forest surrounds the house on all sides, leaving a skylight of blue above us. I brush my thumb on top of Ryan's hand.

"Big day today," I say. He nods slowly and sips from his mug. The steam hugs his face.

"Remember homecoming?" I bait, seeing if I can break his silence.

"I remember that black silk dress," he says, and just the corners of his eyes crinkle into a mischievous grin.

We had stood on this deck at the ripe age of seventeen, posing for photos before our senior-year homecoming dance. My hair was too blonde and my eyebrows too thin, but I stood with the confidence of a girl becoming a woman. I was so secure in my place in the world. He had looked dapper in a suit and a full head of curly brown hair that had not yet been sacrificed to the Marine Corps.

"Are you excited to see everyone?" I ask.

"You know I am." The cicadas ramp up their morning chorus.

"I'm always excited and a little nervous," I admit.

"I know that too. Just be you," he says.

He is right. Today is a day of celebration, but there is pressure. The pressure I put on myself to make it seem like we are okay. Everyone wants to see okay. They do not want to see the pain that throbs between my eyes as I think of all the friends I have just left behind in Morocco. They do not understand that I feel stuck between hello and goodbye. The comfort our blood family can give us is temporary. We have ten days left in America before we are back on a plane. We will head south to Ecuador and figure out how to make its capital, Quito, our next home. We are between worlds, floating between one home and the next.

"I'm going to go try the airline," I say.

"You sure that's a good idea?"

"No." But I get up anyway. I have to hold onto hope that I can find the stroller and the elephant quilt. Evelyn asked for it last night before bed, and my face crumpled. It feels like I've started this next chapter with a failure.

An hour later, I swivel out of the office chair.

"Any luck?" Mom asks me, and I wince.

"Is Ryan still outside?" I ask.

She nods and says, "I've got Evelyn. Go get it out."

When I was little, my dad once accidentally started a grease fire in our kitchen. It rose higher and higher and I remember watching him from underneath our dining room table while flames licked up to the ceiling. I pulled my knees into my chest and stayed there. I felt helpless. He ran to the bedrooms and came back with his arms full of towels. I knew Mom would be so mad about those towels. He used them to beat the fire. He yelled at me to crawl past him and out the door, and I did.

When Mom's car pulled into the driveway, my sister and I were sitting on the porch surrounded by burned towels. She flew from the car to check on us. A small gray stream of smoke rose up from the towels, tiny puffs of ash. She did not even notice. Her eyes flew wildly around until she saw the two of us and Dad, all safe. The four of us hugged for a long time. We had everything we needed.

The porch door bangs shut behind me as I run across the rough boards and out into the bright backyard. I bend over in the grass, hands on my thighs, and dry heave. Ryan sees me and starts running. He is at my side and I'm clutching his hand, my face growing hot with furious tears. I throw a mini-tantrum. I become a small child, trampling the grass with my bare feet and screaming obscenities into the blue sky.

"They can't find it," I cry, and he pulls me in. "They said they'll keep looking, but—" I drop my body down into his chest.

We stand for a long time until I shudder. "It's not really about the stroller."

Ryan smiles. "No."

"But the quilt, I worked so hard on that."

"I know."

"It was her baby blanket."

"I'm sorry," he says.

The porch door swings open and my mother-in-law releases Evelyn out onto the lawn. She runs on strong legs up to us. Her small hands wrap around our knees. I breathe out the first tiny wisp of all there is to let go of. I imagine it like one of the small black smoke tendrils, rising from the ruined towels. I tighten my grip on both Evelyn and Ryan. We have everything we need. We have suitcases, a lost stroller, and more hurdles to jump. But there will be friends and family to hug today.

We are in the place where everything is both lost and found.

The homecoming party goes off without a hitch, and soon we

migrate down to my nana's house an hour away. Now it's time to see the other side of our family. It is a juggle, but so far we have kept all our balls in the air.

When Nana opens her front door, she falls into my arms, a mighty but tiny sparrow. Her puff of white hair smells like it always has, perfumed with roses and soft on my cheek. Her eyes still sparkle like the sun setting across the ocean.

The first night at Nana's, we're getting ready to put Evelyn to bed and my mom fluffs a pillow for her. "She set it all up for you guys," Mom says, gesturing around the basement, which my Nana has gone to great lengths to prepare.

"Did you know this was their first furniture purchase?" Mom says, patting the cream pull-out sofa where Ryan sits.

"Really?" Ryan reaches for my hand, and I pull him up.

"I sat there every Christmas Eve," I tell him and look around at the rough wood paneling. "Not much has changed, really."

The only way in and out of the basement is a heavy wooden sliding barn-style door that creaks and swings a bit when Mom leaves us. My Grandad Gene's paintings line the walls, drawn from either photographs or his own memory. They transport the viewer to his childhood in Maine. Lighthouses, waves crashing against rocks, covered bridges on dirt roads, and lobster traps. One summer, Grandad Gene ordered lobsters from Maine for the whole family. We piled around the table, red-and-white checkered bibs at our necks and butter dripping down our chins. Every wall in this basement whispers to me.

I glance into the next room and see that Evelyn is already dozing on her makeshift bed of cushions. She looks so snug and comfortable, a perfect fit for this nomadic existence.

When we arrived here, she accepted my Nana with a huge hug. Their blue eyes measured each other and filled up with love. At this age, Evelyn delights in whatever her present situation is. She may not think of roots yet, or the lack of them, but I do. Her little chest has breathed in three different homes and she is only three years old.

"Come on." I lead Ryan through the basement. It is like a time machine and I want to take Ryan on a memories tour. We walk into Grandad's office. His large wooden desk is still here, with his name-plate and pen holder. He used to sit up so tall. "He always had a new story to tell me or slides to show us of their travels," I tell Ryan. Ryan picks up a pair of wooden shoes. "Those were from Holland," I say. I used to put them on and imagine myself as a Dutch Dorothy from *The Wizard of Oz*. If I clicked the wood together three times, maybe I could travel across the sea like my grandparents.

Ryan traces a finger across the spines of the *Encyclopædia Britannica*. "This is how people used to see the world." He tells me his parents had a set, and that he used to sit with the volumes for hours. He would flip through the pages and imagine all the different places, people, and animals. I squeeze his arm. We both know how many flights we have stepped off at this point. The world is no longer flat images and text. This knowledge is a gift, but the basement forces me to look back at my life and question my beliefs on where it is headed.

I hear Evelyn sigh from her pile of blankets. I do not regret my insatiable appetite for discovery. I have relied on the next adventure and change. But part of that hunger has always been a desire to know who I am. Becoming a mother broke the image of who I was and remade it into something better. Evelyn's little hands curl into fists and a wave of comfort washes over me. Even with our bags strewn on the floor and another country ahead, I am content. I am no longer chasing a dream. I am living it.

A steady physical house is not in the cards for our family right now, but I can still teach her the value of a home. It is not a place, but people. Home is growing into who I am. Ryan and I walk through the rest of the basement rooms. This house has given me permission to always be myself. It is filled with people who love me. We can provide that no matter where we are.

The barn door groans on its track and Mom slides back in with a rolled-up kitchen towel in her hand. She motions for me to sit down. The pull-out bed squeaks under my weight.

"I'm really sorry about the quilt," she says. My heart clenches. "I know how much work it was." She pats my hand and unwraps the towel.

I see the first glint of bronze and hear a faint jingle.

"It can't replace what you lost, but—" The final flap of the towel falls aside. The elephant mobile Ryan and I purchased long ago in Disney World shimmers, even in the dull light of the basement. Its bells jingle softly as I pick it up and cup the elephants' feet in my hands. A piece of our past to carry into our present. I look up at Mom.

"Thank you," I say.

"It was waiting for you," she answers.

Settling in Again

WE ARE ON ANOTHER PLANE, coming in for a landing after another long flight. The aircraft shudders to a halt and the intercom crackles.

"Ladies and gentlemen, we have arrived. Welcome to Quito, Ecuador!"

It's the beginning of another new adventure for our family.

My daughter's sweaty palm releases mine, her face pale and tired from traveling. The white airline sick bag lies crinkled and unused next to her. I'm thankful for its emptiness. I keep hold of Ryan's hand and our eyes meet in solidarity; it's time to work.

We shuffle around in the seats, slamming spilled items into packs and urging, consoling, and commanding Evelyn to grab her own small suitcase. She whines all the way out of the skinny aisles and collapses into her new stroller outside in the gangway.

What follows is a jumbled routine we know well. We push our myriad items down fluorescent hallways. Stand in the immigration line. Passports stamped, haul onward to baggage claim, where all of our six suitcases tumble down the ramp and into our hands. With six checked bags, three carry-ons, two backpacks, and one just-in-case suit bag, we could be a traveling circus. I'm thankful for each bag as I help heft them onto the cart.

They're all here. We're off to a great start.

My gratitude does not keep the color from rising in my cheeks as I wonder what others might think. I'm embarrassed by the sheer volume of our stuff. This is pride talking, wishing we looked like a rad, travel-savvy crew with only backpacks to our names. I cut this thought off with an inner smile as I remember: We are not here to be tourists in this place. We are here to live.

Our crazy conglomerate of baggage does not weigh us down much save for the brief period of loading it into the van that waits for us. The night air feels crisp and cool on our skin as we pile in. The forty-minute drive winds us into town to our temporary apartment. I can't see much of what Quito looks like in the dark, but I see the layers of twinkling lights that welcome us.

"Mama, look at the fairy lights!" Evelyn says before she gives way to exhaustion and slumps into my lap. The van pulls over on a side street and we tumble out and up into our temporary quarters. It will be another four weeks or more before we are moved into our permanent housing.

I hold onto our most precious baggage, our girl, sleeping in my arms. We ride the elevator up to the apartment. Our motions are practiced and fluid: Determine who sleeps where, set up the monitor, and plug in the sound machine. We decide to leave her in her travel clothes. I peel off her shoes and loosen her French braid. Her silky brown hair falls across the pillow and her small arms wrap for a minute around my neck as she breathes a "thank you" into my ear. Tugging sheets around her, she's set to eke what small hours of sleep she can out of this first night.

Ryan and I locate his work clothes for the next day. It's already past midnight, but he will go to work as early as eight. As our heads hit the pillow, our hands link again.

"We made it," I say with a sigh, and wait for sleep to wash over me. It's a fitful night, as foreign noises surround us. The combination of Quito's high altitude, our jet lag, and the newness of the space wakes both of us incessantly.

In the morning a door cracks open and Evelyn calls out for us frantically. She barely remembers coming into the apartment, let alone where to find us in it. She hears our voices and races to crawl up into our bed. She clings to me, smiling and ready to explore, asking for food, a potty break, and "Where is Bunny?"

Her needs fill the air as I peel myself from the sheets and squint into the light. Our suitcases are chaotically scattered across the floor. We'll be thankful for them time and again as the days stretch to weeks and perhaps even months without the rest of our household goods. Whatever lies in these bags will clothe us, wrap us in comfort, and keep us busy in the days ahead.

~

Any time water bubbles over the side of the pot and touches the surface of the induction stove, the entire unit clicks off. This is one of many things I am learning as we adjust to life in Ecuador. I turn around from chopping vegetables on our one "welcome kit" cutting board just in time to blow the bubbles back down to a non-threatening level. Evelyn peeks her head up above the counter and grabs the boxes of macaroni and cheese.

"Yummy, Mommy!" she says while she shakes them like maracas in her small three-year-old hands.

"Yummy, Mommy," Ryan repeats as he swings through the door of our temporary housing. I grin at him.

"Welcome back. These came all the way from the States," I say.

"Ahh," Ryan says, scooping Evelyn up and kissing her cheeks, "a suitcase specialty."

I nod. "We need a win today."

"How come?" Ryan asks.

I fill him in on our first long day in Ecuador without him. My list rotates back and forth from the positives to the negatives. Evelyn had gotten sick, most likely from the altitude but maybe not. Quito sits more than nine thousand feet above sea level. My lungs and legs reminded me of this adjustment on my search for a minimarket. I secured bread, eggs, milk, cheese, fruit, and cereal, but greens and meat must be hiding in a different part of the city.

I had been able to fit my findings in the bottom of the stroller, and placed a giant water jug in Evelyn's seat so I could push it all back. My biggest challenge of the day had been working to teach Evelyn not to open her mouth in the shower. We aren't supposed to drink any of the tap water.

"The other good news is that the Wi-Fi is great, I've already called Mom and Dad," I say, finishing my account of the day.

"That is a bonus," Ryan says as he swings Evelyn toward me. She giggles.

"Not just a bonus, it's a luxury," I say with laughter in my voice, tickling Evelyn under the chin.

I remember our first six weeks in Morocco. A short seventeen months ago, we had been dropped off in our home there just before Evelyn turned two. I have a photo of her birthday dinner, taken less than a month after we arrived. She has on a green bib and a giant grin. She's sitting at our dining room table with bare walls behind her. We were still waiting on our unaccompanied air shipment, and we had no idea when our household goods would arrive. We were eating macaroni and cheese that day too.

"Are you ready, baby?" I had asked, bringing out our bowls.

"Yeah!" She clapped her hands and swung her legs in her high chair. I placed her bowl in front of her.

"Daddy?" She asked.

"Daddy will be home when he can," I said, squeezing her hand.

I don't remember if he came home in time to eat birthday cupcakes with us. If he did, I don't have a picture of it. I do remember how isolated and sad I felt in those first few months in a new country. I remember my determination to make the large, cold, marble house feel like home. I remember how torn I felt as I wanted to support my spouse while facing the toughest transition I'd gone through as a military spouse. The cupcake mix and two boxes of white cheddar macaroni I had brought in my suitcase were a precious reprieve from all I did not understand and had not yet embraced about our current situation.

The timer beeps in the kitchen and my mind returns to the present. All three of us gather around the sink as I drop the hot noodles into a colander. Evelyn waves her hand through the steam. Ryan carries her over to the stovetop and she helps us drop butter into the pan, then with milk and cheddar powder. I stir it all together as Ryan puts Evelyn in her booster seat at the dining room table.

As I portion dinner into bowls, I notice that the noodles look small. I pick one up, squeeze it between my fingers, and then pop it into my mouth.

"Ryan?" I say with concern. He looks up. "Try this," I say, spooning some into his mouth.

He tastes the noodles. "It's ... " he hesitates.

"Crunchy," I finish his statement for him. "I set the timer," I say and check the box.

"It's the altitude," he says and explains that at higher elevations the water boils faster but at a lower temperature. I stare at him. This must be some part of science class I missed.

"Oh," is all I can say. My shoulders sag.

We set the bowls down at the table and crunch our way through the noodles. I am so disappointed. My "win" for the day, macaroni and cheese hauled all the way to South America, is a complete fail.

Just then, Evelyn smiles from her chair and says, "Yummy, Mommy!"

Ryan touches my foot under the table with his and I catch his eye. We both start laughing. Not loudly at first, but a slow, quiet session of chuckles that grow into guffaws, then shoulder-heaving, tear-inducing, snorting peals of laughter.

After tucking Evelyn in, Ryan and I lie down too, facing each other on our pillows. We talk through what else we can do in this holding period. We still have a long wait for our household goods and a house to put them in. We'll make a trip together to a larger grocery store tomorrow in hopes of finding more of what we need to fill our pantry. There is an embassy party coming up soon, so I hope Evelyn and I can make some friends.

The room grows quiet. Ryan closes his eyes.

"Just so crunchy," I murmur, and he opens one eye. We fall into fits of laughter again.

It was a terrible dinner, but I'm grateful. Our past experiences have helped me grin my way through this round of transition. I won't laugh at everything, but I can find more humor in the unknown, instead of the pain and impatience I felt in Morocco. I close my eyes and reach for Ryan's hand. I marvel at the ridiculous wonder of attempting to settle in as foreigners to a foreign land. But we are doing it together.

In the morning, I order a twelve-pack of macaroni and cheese from Amazon. I eventually figure out how to make the perfect high-altitude mac and cheese. In our next eighteen months in Ecuador, I never put a cook timer on again for noodles. I learn to test the noodles by sight and taste. Sometimes I have to feel things out for a while before I settle in.

Leftover Friends

THE SPRAWLING LAWN of the ambassador's home in Quito is sprinkled with people wearing red, white, and blue. The aroma of grilled meat flavors the wind. A big tent holds lines of tables, and a band plays festive music. There is a bouncy house featuring two giant slides set up in a corner of the lawn, and as soon as Evelyn sees it, she tugs my hand toward the noise of shrieking and laughing kids. I make eye contact with Ryan before I get pulled away, one raised eyebrow that reminds him of what we discussed in the car. I'll take the first shift of kid play, but then he's on.

While he's at work, he's been able to get to know this group of people. I've been apart from those relationships, managing the home front, which means reading, playing, and spending the days on my own with Evelyn while we wait to move from our temporary apartment in the middle of central Quito to a home the embassy housing staff is preparing for us. We were told it would be ready in ten days, but that has extended to more than thirty. It's frustrating, but there's not much we can do except practice patience.

The first month downtown passed fairly quickly. We have toys, books, and crafts I had brought in our suitcases. Evelyn and I break up our time by alternating trips to the nearby park, which has multiple engaging playgrounds, pedal boats, a botanical garden, and plenty of tall trees and bushes for building fairy house towns beneath. I'm becoming quite adept at rigging up wee log cabins decorated with grass and flowers. I've done the majority of this waiting time in good spirits, but those are starting to wane. I long for a space to call our own, a car to give me mobility, a routine that settles our hearts into nurturing normalcy, and friends to populate this new life.

When I met with the community liaison officer at the embassy last week, she shared some books from her office library and asked if we'd connected with anyone yet. I told her we were doing all right, but Evelyn and I were both eager to make friends. As we talked, she discovered we were considering homeschooling Evelyn for preschool, and she told me about another homeschooling mama in the embassy community who also enjoys writing. Her name is Kristin, and I'm hoping she is at the party today.

At the bounce house, I talk with some of the other parents. We share the normal banter—where we've lived, what we do or don't do for work, our kids' ages, and so on. They bring warmth and kindness, but there's no Kristin yet. I have a sinking feeling I won't get to meet her today, and I realize how much I've been counting on that to spark my first friendship here in Ecuador.

I know by now it only takes one person to see my heart—and me theirs—for life to feel substantially more supported when I'm navigating a new place. Cultivating a new home and routine takes time, and I have patience for that. But the need for building community has a special urgency for me. I am in a hurry to put that piece into place, because I don't feel at home without relationships.

After Evelyn and I have spent thirty minutes or so at the bounce house, Ryan waves to me from across the lawn. I grab Evelyn's hand and we make our way to him.

"How's it going?" he asks, and I answer, "It's going."

"You seem disappointed," he says, scooping up Evelyn.

"I'm not, I'm just—I thought I'd meet that other mom, Kristin? I haven't found her yet."

Ryan says he's not sure who she is either, but suggests we go get some food. As we head toward the food tent, I see a woman making her way across the grass toward me. She has brilliant blue eyes, blonde hair, and striking red lipstick. I try to calmly walk toward her, but it feels more like the skip-bounce-run of an overeager giraffe, a tangle of legs and long neck extending to make contact.

"Are you Kristin?!" I say, tentatively reaching out a hand toward her.

She takes a little step back but responds with wide eyes and a big smile.

"I am," she says.

I close the distance between us and shake her hand. I gush about how I was told to find her today and that the CLO thought we would get along. It tumbles out in a span of less than ten seconds.

As Kristin and I shake hands, I am hopeful she'll be a person who understands my excitement. We chat through various pieces of our lives, and I see Ryan smile at me from across the lawn. I'm past the first hurdle. He gets us food and drink, and we end up sitting in the grass together with our families, talking. I learn Kristin has a daughter a few months younger than Evelyn, and she offers to come pick us up for a playdate. We exchange numbers and I leave the party with the flickering feeling that this could be something.

Months later, Kristin and I are in her cozy living room again, chuckling about how I approached her with so much energy the first time we met. Our daughters spend afternoons playing together while we unpack favorite books, writing material, and thoughts about maybe homeschooling for the long term. We cheer each other on in our next steps as creatives too. She begins her master's program as I start working on a book idea with a writing mentor.

Within a month of arrival, I'm thankful I've found that one key friend, the one who unlocks a sense of community in a new place. But I'm also aware that Kristin leaves eight months before we do, and a little alarm bell pings in my brain to keep my eyes and heart open to more friendships. It's not that one friend is not enough, but I can't build my entire sense of security on one friendship. It's a lot of pressure and expectation on one person.

In this transient life, the coming and going of friends is constant. It feels like I'm saying hello and goodbye on repeat, sometimes in the same breath. It is painful and yet purpose-filled, teaching me to hold the bonds of friendship with open hands.

As Kristin and I go deeper in our friendship, we dive in quickly, trusting one another to hold each other's hardships, joys, and truths. It is sometimes raw and often beautiful. Like divers exploring oceanic zones, we submerge each time we meet up, allowing our friendship to expand in each layer, from the sunlit, easy zone of basic information to the twilight depths, where we tap some of our long-held emotions. With those levels complete, we hold our breaths, and we sink down to the trenches of our experience.

The deeper we go, the more the risk, the more we hope the other person continues to meet us there. With depth comes the thread that ties one to another, expanding each heart with the understanding we'll try to be friends after we part. I wonder if I would give my heart so quickly if our time weren't so limited. It is awesome to dive so deeply but also dizzying. Friendships like these buoy me up. I rise and fall on them.

We finally move into our home in midsummer. This time it is a sixth-story apartment building with a charming rooftop garden and amazing views of the Andes mountains. Our air shipment arrives, adorning our beds in familiar quilts, drying our bodies with fluffy towels, and filling our cupboards with a familiar array of cookware. It takes a full five months to receive the rest of our household goods, and six for our car to arrive. Until then, we remain searching for the familiarity of schedule and friends.

On a bus ride through the winding back roads of the Ecuadorian countryside, I first notice Rebecca and her husband, Joshua.

They have five children spanning ten years old to one. One of their little girls, wearing a pink hat, looks Evelyn-sized. While our family enjoys the view and the roads that curve in and out, Rebecca and Joshua juggle vomit bags as their kids suffer repeatedly from motion sickness. Ryan and I admire how they handle their kids with patience and care. We offer encouraging words and talk to them intermittently about their lives and how they're settling into Quito.

When the bus comes to a halt at our destination, a rose farm, we all tumble down to stretch our legs and walk through the beautiful greenhouses. I see color return to Rebecca's kids' faces. Evelyn teams up with the other kids to carefully pluck an assortment and abundance of rose petals off the ground. They gather them by the handful, scooping them up into piles of blood red, blushing pink, and butter yellow. Rebecca and I have time to exchange information and make first-date plans as the kids collect petals.

Back at home, I place the velvety soft rose petals in a jar by the window to dry. Roses grow wonderfully in Ecuador, where the strong equatorial sun pulls their stems skyward, making them straight, tall, and elegant. Yet, a dozen of these regal flowers can be purchased for less than five dollars.

Soon, Kristin, Rebecca, and I become a regular trio. We gather for hot *horchata* tea in our hectic living rooms and enjoy lively discussions while our kids play. We cheer on one another through a sweaty group workout, push our little girls on the playground swings under bright blue skies, and attend intriguing book clubs. Rebecca opens her home for a Bible study, and I sit at the feet of these women, learning and growing. Months into our time in Quito, I am at ease, and I realize it is because of how quickly my friendships here have opened and formed.

The rose petals I placed in my window curl and crisp in the sunlight. One day I empty them, and Evelyn plays with the petals on the dining room table, arranging them into a circle of color.

They have retained their vibrant colors and a hint of sweet fragrance. Time has transformed their velvet texture into a tougher husk, but they are still delicate to the touch. I think of the beginning blooms of friendship with Kristin and Rebecca, and of how being with them has brought a new season of substance and sustenance to my life as a woman, a friend, a mother, and a spouse.

All too soon the season turns, and our three amigas fade to two as Kristin and her family must move onward to their new assignment in Grenada. While her departure does not come as a surprise, it is still painful to walk to the park and feel her absence. I stroll up her street one day and hear a new family playing in the backyard behind her garden wall. Evelyn asks to go in and play.

I wonder about Evelyn's memories of this place and its people. I hope somewhere in her body she retains the way we felt safe and loved here, nurtured and seen, so that in the future she'll seek and create that kind of community again. I remind her that it is no longer Miss Kristin's home, and we both need a hug after that. We walk onward to Rebecca's home.

Rebecca and I create a routine of trading preschool instruction days with our girls. After selecting a manageable curriculum together, we build upon the girls' wonder and skill. We then have lunch as a group and allow the girls some unstructured playtime. We help each other clean our homes and do dishes, and it is in these mundane moments that our relationship moves deeper.

One day, after helping her mop the floor at her house, I confide, "I think this is my favorite kind of friendship."

"What do you mean?" she asks, pulling a plate of leftovers from the microwave for our shared lunch.

"The ease of doing life along with someone else," I say. And what I'm trying to put into words is that between the two of us, we feel no need to impress. There's a special softness that comes from being together in this stage of little kids in this time and place, encouraging one another through everyday tasks.

"I had this with one other person before," she tells me. "We called each other 'leftover friends,' because we were comfortable enough to eat one another's leftovers from the fridge together."

I smiled and nodded. "I love that," I say, realizing we have reached that kinship when I can freely open her cupboard for a teacup or dig through her fridge for a snack. We go about our days supporting each other. Our girls delight in their friendships. And we end up exchanging a fair amount of leftover food for our lunch, each of us enjoying the other's effort, offered up in generous spirit.

We know this time together, our routine and what we've built, will end. My departure date is set just as Kristin's was. But until we reach our goodbye, we're all in. Our hearts can settle for the time we have in a friendship such as this.

Movable Roots

"WHERE IS HOME for you?"

The woman in the chair next to us has been watching us. When I first saw her silvery hair, bright scarf, and warm eyes, I could tell she was an airport talker. She reminded me of my mom, who can make conversation with a rock. But every time the woman wanted to connect, we have been occupied by reading stories or eating donuts. Ryan takes Evelyn to wash the glaze off her hands, giving the stranger a window of opportunity for her question.

"Oh, well, all over really," I hesitantly answer.

When I was little, according to my mom, I would crawl up into her lap and hold her face in my hands. The hunter-green rocking chair would sway us back and forth, and I would say, "I'm never leaving, Mom. I'm always going to be right here."

She would laugh and say, "You don't know that. Wait and see."

I look at the woman and try another answer "We're currently living in Ecuador—that's where we're headed today." I cross my arms over my chest.

It's hard to explain. We're returning from a trip to Peru. But I don't know if I can say we're going home. Quito is another temporary chapter in a string of locations around the globe. Each place has been special in its own way. We have found great friends who have become our community. Our parents are always glad to traverse our exotic locations. When they visit, they grant us a break in resident status, allowing us to be tourists. In every place we live, we choose to make our time worth the painful effort it takes to pack up and start over again and again.

When we land back in Quito, we'll drive our little blue Toyota down the curving mountain roads, past signs with Spanish names, and into our small gated neighborhood. In our sixth-floor apartment, I'll unpack and wash laundry, refill the fridge, and check email. Life will hum on in Ecuador as it has for us in northern California, South Korea, Hawaii, Morocco, and wherever else the United States Marine Corps will send us.

The woman is still waiting patiently for my response. I stumble on, toying with the zipper on my maroon jacket. "I mean—my husband and I? We're from Ohio originally. But that was a long time ago. Now we like to say we are each other's home."

"That's nice," she replies, settling back in her chair. "Where else have you lived?"

I prepare myself for my go-to speech. Straightening my shoulders I deliver all of our locations and include, "I even worked on a cruise ship!" I love it when eyebrows go up. Our frequent moves have made us unique. I revel in the distinction.

I pause with a little bravado, pulling my fuzzy travel hair up into a top knot, and set up the laugh line: "Yeah, I always say that getting married didn't scare me. Having a baby didn't scare me. Moving all the time doesn't scare me—but choosing one forever home? That's terrifying!"

My silver-haired airport comrade laughs on cue. I study her face to see if she knows I'm not being completely truthful. I am scared of moving all the time, because I am frightened of what it's doing to my heart. Instead of seeking soil for my roots, I have mutated into an air plant, adapting to the environment and taking sustenance where I find it. I soak up the same sun in different spaces. I blossom for a season, then hibernate during change, ready to flourish again elsewhere.

Evelyn flutters back to hold onto the side of my leg. When I look down at her, she's smiling, unfazed by another airport day. As our flight is called, she pulls on her backpack and waltzes right up to the line. Our brown-haired girl walks with spirit, confidence, and better travel savvy than many five times her age. This trip to Peru was her tenth country so far. At age four, she has yet to live in the continental United States. Her passport says "American," but I don't know if we have taught her what that means. Do I remember what it means to be an American in America?

Two days before we left on this trip, I attended a seminar at the embassy called "Finding a Place in this World: Third Culture Kids." In the dim, cold conference room, we parents sat on the edge of our seats. When the presenter walked in, flicking on the projector, she shook us from our nervous silence.

"You have not ruined your children," she said. A tittering of quiet laughter spread through the dark, and she added, "But we do have some ground to cover today."

For the next hour, we dove into how working, living, and moving to multiple locations shape our lives and the lives of our children. The list of positives includes the opportunity for travel, acknowledgment of more than one language, diversity of thought, and a broader worldview. On the other hand, we have to understand that we and our children are forever changed. They have not and may never associate home with a place. We might not either. We have to teach them, and ourselves, to say goodbye, let go, move on, and start over.

I imagine roots again, sliced and transplanted, and wonder if we can learn to settle, or if we will forever seek the comfort of change.

It's time to board, and Ryan guides us into the line.

"Have a good trip!" the woman calls to me, smiling and waving.

I smile and wave back, but my mind catches on the word "trip." We're done with the trip. We are going home. Home to Quito, if home means an eighteen-month cultural immersion before another set of orders is delivered and we're whisked off to another corner of the globe.

I take my seat between Ryan and Evelyn. He is quickly immersed in a crossword puzzle, and I help her set up a movie on her tablet. I close my eyes, and our plane lifts off the ground. Sandwiched between these two people I love, I notice the simplicity of Evelyn's small folded hands in her lap, Ryan tapping the tip of his pen to his lips, and the way we naturally fold in toward each other.

After years of transience, our roots have become embedded in the people around us. Our relationships are walking homes where we hold one another gently inside.

As I feel the familiar sensation of the plane beginning to level off, taking us from one place to another, I reach under my seat to pull my journal and pen out of my backpack. I click open the black pen and spread stories across the page, scrolling ink across lines. The act itself is comforting, the heft of the book warm on my lap.

Whether I chose this sojourner life or it chose me, I have a heart that carries many places and people. They live on my person, branching out and extending my tendrils. I might never see some of the houses I made into homes ever again, or hug the people who loved me in those places, so perhaps the purpose of these threadlike appendages is that they are meant to stretch out and find a support of their own. They can root as stories, built from the mortar of memories, a growing roof that extends across time and space. Our stories are our homes; we can take them with us.

Renewed Dreams

MY LEFT KNEE JIGGLES my right up and down, up and down, a quick cadence of nerves. I am at the embassy in Quito in a room on one of the upper floors, waiting for my meeting with a career counselor. The room is a mix of creamy whites, chrome, and pops of color. I look down at my nervous leg and bouncing foot, clad in one of my remaining pairs of professional-esque black pants and black ballet flats. There is no reason to be nervous. The career counselor I'm seeing today travels from post to post meeting with spouses like me. She works for the Global Employment Initiative, a US State Department program to help family members with career challenges while living overseas. I am not interviewing for a job. I am here to talk through my past jobs and explore what I might want to do next.

Bounce, bounce, bounce. I start to count my heel drops, breathing in through my nose and out through my mouth. I check the folder in my hands to see that the resume I updated for our meeting is still in place, with all its odd information. Dance position here, writing job there, freelance pieces scattered all over. The years blend from one post to the next, filled by whatever was available for me to do. Then there are the gap years in which I've been raising Evelyn, sometimes with Ryan home—often with him not.

I glance down at last year's updates and wonder what it would have been like to come to this meeting with the title "Community Liaison Officer" on my resume. *Would I feel more comfortable?*

Would I be more confident? Would our bank account be further along? I cannot answer those questions, because in Morocco when I was asked if I'd like to apply for the position, I declined.

I'd like to think I gave it a night's sleep, but I know the pattern my mind likely followed. *Will this position require placing Evelyn with a nanny or childcare all day? Yes. Could this deplete the time Ryan and I can devote to healing our marriage? Yes. Will we be having another child? Maybe not. Maybe Evelyn is it. And I want this time with her.*

That sealed the deal. I said no to the opportunity to add that to my resume. Ryan and I adjusted our finances so I could stay home, which we knew we were fortunate to do. I took an odd freelance writing piece or two and enjoyed sharing my dance experience with the Crescendo program, and that was enough for us.

The newest line on my resume is a position here in Quito as the editor of the family newsletter. When I first heard about the job opening, a little beacon blinked in my heart. I asked Ryan to take Evelyn for a weekend while I locked myself away in our home office to design my application. I made a twenty-page "audition" newsletter and emailed in my submission. It felt exciting and right and also interesting that my hours of fine-tuning my writing and learning the basics of graphic design now opened me to this potential opportunity.

I got the job.

I work from home, which is nice and also a new transition as Evelyn learns how to be comfortable with more independent play while I type away on the computer. I learned how to set up stations for her like Play-Doh and rice bins, audiobooks for her to listen to as she plays, and a timer for myself. I also learned how to surrender to the television as a tool every once in a while, which Evelyn doesn't mind one bit. We're still doing preschool from home, with plenty of time for adventures, reading aloud, and playing with friends. It has worked out, and it has made me curious about what else I might want to do. So here I am at this meeting to explore the possibilities.

Bounce, bounce, bounce. I set my folder down and breathe deeply in and out again.

"Mrs. Swoboda?" My knee stops. I place both feet on the floor and stand to greet the career counselor, who welcomes me with a warm handshake. She leads me to a long rectangular table and informs me that this meeting will focus on questions I have about work transitions, extra schooling I might be interested in, and remote positions, as well as looking at my resume and seeing what we have to work with.

"Is that what you have there?" she asks, nodding at my folder.

"Yes!" I slide it across the table and sit back in my chair. I cross my legs, and the bounce, bounce, bounce comes back. Her eyes look carefully through each year and position I have listed. I try to read her face, but it is open, without flickers of concern.

"You've done a lot of interesting things," she says as she flips the paper over.

"I have. That's true," I say. "But it's such a mix. I went ahead and listed everything I could think of between writing and dance but—" I pause, searching for the best word. My leg moves to a jiggle again, in place of a steady rhythm. "But it's all a bit cobbled together, isn't it?" I lean forward and place my hands on the document. I explain all the transitions we've had in recent years, and how we've been focused on our family as well.

"Can I say something?" she asks.

"Please," I say. I can feel myself filling the space with words on top of words to cover my nerves.

"You haven't been cobbling it together."

"What?" I ask, and my knee comes to a pause under the table.

"You said the word 'cobbling' about your career moves. That makes me think of someone slapping things together. But what I see here in front of me, is someone who is being very intentional with their time."

She slides the paper back to me, and I look down at it. And I nod slowly.

"Thank you for saying that. I don't think I've ever looked at it that way before."

"You have a lot of skills that have been developed because of this lifestyle. I see someone who has creative strength, adaptability, innovative ideas, and the ability to pivot when needed."

I place both feet beneath me and nod again, pressing my lips together and exhaling through my nose. She goes on to tell me that there's so much I can do with my resume and skill set. I can go on a writing track, back to dance, or choose a whole new direction.

"And it's also okay to not know—"

"What I want to be when I grow up?" I ask and we both chuckle.

"Yes, we're all trying to figure that out. You're going to keep changing. And then when it is time, you'll begin to pursue the next challenging and interesting opportunity for you. But until then, please do not diminish what you've done thus far."

I can feel a warmth in my chest. Her words are like a calming hug, reassuring and reminding me that what is ahead can be just as exciting as what is behind me. We end our session by talking about some directions I might want to go in and how to utilize some of the different online resources available to me through the global employment program.

I leave feeling rejuvenated. This step of becoming the newsletter editor at the embassy has given me confidence. I enjoy this kind of creative, organizational work that helps bring people together with offerings and reminders about community events. What can I dream next?

At home that night, I sit with Ryan on our couch, my legs resting across his lap, and unpack my session with the career counselor.

I am reminded of our life in Monterey, and how he helped me transition into a new dream of teaching dance.

"Do you miss teaching?" he asks me.

I meet his eyes. "You know what? Not the way I used to. It's like I know there will be a time for that again, but it's okay that it isn't now." He pats my thigh. It doesn't feel like a grief like it used to, I explain, but a goodness, a chance for me to hone some new skills and to be home with Evelyn.

"I was thinking I'd like to take some writing courses," I add.

He squeezes my leg. "That sounds like a great idea."

"Do you think there's time?" I ask. He's also taking courses to finish his bachelor's degree.

"I think we make the time," he says.

I smile. "Okay. Now tell me about you."

Ryan says he's feeling pretty good at work. This posting has had a whole different vibe than Morocco. He feels more confident about what and how he needs to run things. He's also excited about his upcoming promotion.

"Finally!" I interject, and he grimaces.

"I know. I know." It's my turn to pat his leg, and he grabs my hand to squeeze it. "That was rough," he says.

"Yes, but it also taught me a lot," I say and he kisses my hand. "Tell me about that," he says and I accept his invitation to go back to this raw space together.

I close my eyes and picture the day in Morocco when we found out he didn't get promoted to gunnery sergeant. I remember scrolling through the names, looking for ours, but when I slowed down where his name should be, he wasn't listed. Why wasn't he listed? I checked, double-checked, and triple-checked. He wasn't there.

I was in the kitchen at the time, and I looked over the top of my phone at the sparkly scrapbook-style page taped to our fridge. It had our goals for the year ahead. Emblazoned across the top were the words in all caps: MONEY + GUNNY. It seemed silly now, but it represented serious goals for us.

We'd been aiming to save our finances so we could steward well for our dream of buying a home one day. Making gunny would have meant getting a pay raise. Aiming toward this goal, we both tried to look past the long hours he was working, sometimes seven days a week. We tried to stay focused on our mission: do what was needed for his promotion.

When he had come home that day, I didn't mention it and neither did he while Evelyn was awake. But after we got her down to sleep, I found him in the bathroom, slowly brushing his teeth. Our eyes locked in the mirror over the sink.

"I know you know," he said, his teeth gritted. He dipped to spit into the sink.

I nodded, my hands shaky as I touched his back. "Why?" I asked. "Why not you?"

He kept brushing in slow circles as my thoughts tumbled, searching for an answer and finding none.

"I can't understand it," I said. "We've done everything that's been asked of us. You are a good person."

"That doesn't always mean it works out," he said. "It doesn't always mean I'm a good Marine."

He turned to leave the bathroom and I caught his arm. He told me that while he looked all right, that was not how he felt. He wasn't sure how to process all of this and maybe he wasn't ready to hear how I felt about it too.

I slumped against the bathroom door and folded my arms. I wanted to be present for him, to offer comfort, but I didn't know how. I was disappointed, but not in him. I was disappointed in the system he served. None of it seemed fair. Maybe it was a childish response, but I had been witness to his dedication, his sacrifice. Our sacrifice. The time and energy poured into his career made this disappointment sting like a slap. It felt as if we weren't good enough. As a family, we endured the deployments. We took the overseas assignments. We didn't complain. I thought we were checking all the boxes.

He went to bed before me that night. He lay stretched out, arms pulled in. I kissed him and let myself remember.

I remembered him at seventeen, his curly brown hair and kind eyes, pulling me toward him with a laugh. I remembered him holding our infant girl in the hospital room, his bare chest pressed to hers, his love expanding with mine. I remembered that the uniform hanging in his closet—representing the commitment we'd said yes to as a family—is secondary to the family we are building. Ryan is a Marine, yes. But he is, and will always be, himself.

I open my eyes and squeeze Ryan's hand again.

"I did not fall in love with you because you are in the service," I say. "That is secondary to us. I fell in love with you because you are you."

We hold one another's gaze and touch foreheads. I tell him he deserves every good thing, and he always will.

He did receive his promotion the next time around.

We pin him for gunny on a summer day on the embassy lawn in Quito. It is exciting and well-earned, our appreciation enhanced by the delay we experienced. I love taking the photos and eating the cake. But in my mind, where we go from here has shifted. It's not all about career progression, for me or for Ryan. Promotions will come and go. Opportunities may open and close. But we stay.

I stand next to him that day, and I stand with him always. We have many more dreams to realize or let go of, together.

The Mommy Boat

THE CAPTAIN OF THE SMALL GLASS-BOTTOM BOAT cuts the motor and we glide up next to the dock. We are on our final adventure in South America: the Galápagos Islands. It is also perhaps our last overseas adventure for some time, as we have received our next orders to Norfolk, Virginia.

I remember when we were trying to figure out where to go next, over a conversation at the breakfast table about five months ago. We told Evelyn while she was eating her eggs and toast that we would be speaking that morning about our future duty station.

Only four years old, she perked up and asked, "Do I get a say?"

Ryan and I looked at each other in surprise.

"Well, we will certainly listen to your opinion," he said.

"But we will go wherever the Marine Corps sends us," I added. I did not want to squelch her enthusiasm, but I did want to keep her expectations in check. Ryan brought out the list we had narrowed down. We started discussing the merits of Japan and Virginia. Evelyn raised her fork in the air.

"Go ahead," Ryan said, giving her the floor.

"Japan is another country, right?" she asked.

"Yes, that's right," I said.

"But Virginia is in the States," Evelyn said.

"It is," Ryan said.

"And my grandparents are in the States?" she asked, spearing eggs with her fork.

Ryan and I caught one another's eyes again and smiled.

"They are."

"Then I would like to go there," she said with a shrug. Though it was not the opinion of a four-year-old that set us on a path back to the United States, I am glad she weighed in. When I think of being a drive or a short flight from our families, my heart expands and contracts. I dare not entertain the idea, though, in case I jinx it.

The captain moors the boat, and I reach my hand out to help Evelyn disembark. We walk from the dock across rough black volcanic rock and more wooden planks to reach a lagoon. The captain, also our guide, tells us we can snorkel here.

The water is clear and more still than the open ocean, and we will be able to see a variety of ocean life as we swim from end to end. Standing at the ladder where we will climb in, I look into the depths that stretch far below us in the water. Evelyn pulls on her life vest and looks at me.

"Are you ready for this?" I ask, and she nods, but her eyes flash with concern as she steps off the ladder into the water.

Her legs kick, kick, kick in the water and she turns herself back around to grab the ladder. Her pink water shoes bob up in front of her as she clings to the metal rungs.

"I don't want to go out," she says, her eyes welling with fear. We have practiced for months for this trip, attending a lengthy summer swim camp and weekly lessons to help her gain confidence and skill. But here in this lagoon, without a secure pool floor and walls around us, her resolve sinks to the sandy floor. I pull on my snorkel mask. I push away and swim a few strokes then return to Evelyn.

"You can put your arms around me and I can paddle us around in the lagoon. Would that be okay?" I'm asking Evelyn, but I'm looking at Ryan.

"I think that would be great," he answers. I swing my body around so Evelyn can get on my back. Once she has hold of me, I put the snorkel tube in my mouth and place my face down into the water. We're off.

"Are you feeling all right?" I ask her through the tube, as I pull us away from the ladder. I hear a muffled reply from above. "Talk louder so I can hear you," I say.

"I'm good!" she calls from the surface. I turn left and lead us to a shelf in the rocky side of the lagoon. I guide her from my back and inch myself onto it. I help her pull her goggles on and instruct her to keep her feet on the shelf, so she can crouch down and look into the water. She holds her breath and puts her face in the water. When she pops up she is grinning.

"There are so many fish down there!" she says, clapping.

"I know, it is beautiful, isn't it?" I ask, and she nods. "Do you want to go out a little farther?"

She says she would like that. I step off the ledge and prepare myself to take her back out. Ryan swims up next to us and helps her on. He kisses my cheek before I set off again. I imagine it is a silent "thank you," as I am the more confident fish of the family.

Evelyn and I move from rock ledge to rock ledge, stopping and peering down, her face always emerging from the water with a toothy grin, describing the aquatic creatures she has spotted.

We are at the far end of the lagoon when I see the captain wave.

"Time to go," I say. "Climb aboard!"

"All aboard Mommy boat!" Evelyn says, and I laugh. "We're not stopping this time, so keep hold of my shoulders and help me by kicking your feet behind you.

"Okay." She pulls her goggles down over her eyes with determination. I push off the rock and can feel her small hands locked onto my shoulders. Her life vest helps take her weight off me, and I can feel her body wiggling with kicks. I breathe in and out of the snorkel tube, pulling us back to the ladder. It is a longer swim than I remembered. With our stops along the way, I hadn't realized how far out we were. My legs burn and my neck tenses as I work to keep Evelyn balanced safely on my back.

"We're almost there," I say through the tube, willing myself to push through the fatigue. I give a great pull with my arms, but my head goes under. Water floods my snorkel tube, my mouth, my nose. I feel Evelyn tense above me as I try to blow the water up and out with a raspy gurgle. My face is still underwater, and I can see the ladder about seven feet away. I reach and hold Evelyn up in the water, forgetting in the confusion that she has a life vest on. All I want is for her to stay up and out of danger while my lungs feel like they're shriveling in my chest.

The captain is there, reaching out to Evelyn. I push her body out away from me and toward him. When he grabs her, I pop up above the surface, sputtering and coughing.

"Thank you," I choke out and swim toward his hand. He drags me up to the ladder.

I climb up and see that Ryan has Evelyn wrapped in a towel. I grab one myself and pat dry. The whole ordeal might have only been seconds, a minute at most, but my body is quivering.

"I'm sorry, I got water in my tube and—"

"And we're all alright," he finishes. "She had her life jacket on, she was fine. You're okay, right, squirt?"

Evelyn smiles and nods and I kiss the top of her head.

"Mommy boat sprung a leak!" I say and we all laugh.

Offering my child the wonders of the world is both exhilarating and unnerving. I don't want fear to keep me—or my daughter—from a grand adventure when life offers it up.

We stop at one more snorkeling spot before the tour is over. This one is a beach, and I swim out alone to explore a cove. Looking down through my mask, I see one, two, three, four sea turtles grazing below me. I sprint-swim back to the shore.

"C'mon, Evelyn!" I shout. "The turtles! They're here!"

She grabs my hand and we swim back out. I talk to her through the snorkel tube again.

"There's one right below us. Are your goggles on?"

"Yes, Mama!" She shouts from behind me. I lead her around and she floats beside me. I hold her beneath her belly, pulling the lifejacket away from her throat so she can look down.

"On three!" I count, and together we dip our faces under.

One turtle's shell looks almost as big as the top of our Toyota. It bobs around a rock and looks like it is nibbling green matter from the side. It glides effortlessly over to another rock, eyes open and round beneath the blue-green water. I pop my face up to look at Evelyn. "Did you see it?"

"I did!" she squeals. We swim back to the beach, and I sit next to Ryan, digging my toes in the sand. Evelyn runs across the beach, digs holes, and searches for sea treasures. I remember the afternoon back in Morocco where we had built a tent of blankets in the backyard to read books. The move to Ecuador was not far off then, and I showed her videos of Quito and the Galápagos. She loved the idea of swimming near the sea turtles. And now we have.

I pat Ryan's arm. This is our last big trip before we begin trying again for another baby. While I always imagined we would have children closer together in age, that is not how our story panned out. We decided we would wait until we were closer to the end of Ryan's time with the Marine Security Guard. Living overseas, often far from US military installations, complicates the process of welcoming a new member to the family.

As we researched and asked questions, we discovered stipulations if I became pregnant in Ecuador. There was a chance Evelyn and I would need to leave so I could deliver the baby stateside. After weathering one pregnancy spent apart during deployment, we didn't want to choose that again. Also, with the prevalence of the Zika virus in Quito, if I become pregnant here, my ability to travel would be strictly limited. It did not feel like a win-win situation from any angle. We ultimately decided to strengthen our bonds as a family of three while living there.

We also knew we had to heal what felt broken in our marriage before expanding our family. Not a bandage sort of mending, but the deep soul work of understanding who we are again as a team and individually. I think we are there. While I have been patient, the longing for a new baby is becoming more urgent. I lean into Ryan as I watch Evelyn scoop sand into a pile. Perhaps we will all have new dreams on our way to Virginia.

The Fire

WHEN I TURN ONTO THE MAIN STREET leading back to our apartment in Quito, I can feel something is wrong. There are many more people out on the street than usual, some rolling what look like metal trash cans. In one place, a can is turned upright and fire spills from its top, smoke curling into the air. I consider returning to my friend's house, but there is nowhere to turn the car around, and I cannot risk getting boxed in out here.

I drive the short few blocks home, and as I step into our apartment my phone rings. It is Ryan.

"Where are you?" he asks.

"I just got home."

"From where?"

"It's co-op day, and Rebecca has the girls. I dropped off Evelyn."

"I need you to stay put now," he says, his voice holding the calm readiness I recognize as both his professional training and his personality. I also recognize something is up.

"What?" I ask, sliding down to sit on the sofa.

Ryan explains there is civil unrest in the city, stoked by opposition to policy changes by the Ecuadorian government. The American embassy community is being asked to stay where they are until the situation is assessed and resolved.

"I didn't mean to leave her," I say.

"Honey, it's okay. She's in a safe place with people who love her," Ryan says. While I know this is true, all I want to do is run

through the streets and scale the protective fences to reach my daughter and gather her in my arms.

"I didn't know," I say, sinking deeper into the sofa and rubbing my temples.

"You couldn't have. Stay put. We'll figure out a good time to go get her when this clears up," he says.

When he hangs up I look at my hands. They are trembling. I did know, didn't I? I saw the people, the barrels, the fire. I was supposed to come back home today and write while Evelyn is with Rebecca and her girls. Instead, I pace the floor of our apartment and look out the window no fewer than twenty-five times in the first half hour. Our street looks normal.

Ryan gets home late that evening, and the first thing we do is jump in the car to collect Evelyn. The streets are dark. My body is on high alert, watching for elements out of place on the drive, but I keep my tension in check. When Rebecca opens the door, I hug her.

"Thank you," I say.

"They had a great day," she says, and I know this is true. Rebecca's daughter is Evelyn's best friend.

"They had no idea anything else was going on," Rebecca says, while I chew the inside of my left cheek. I know she kept them busy, fed, and safe, even while she must have been feeling the tension of the day. I wonder if I would have the same strength. I hug Rebecca again on our way out, and we discuss how we'll have to take the co-op preschool and playdates by ear until things calm down.

Evelyn is beaming from her wonderful day and is thrilled to be up so late past her bedtime. She tells us about the lesson they had today on *The Very Hungry Caterpillar* and shows us the macaroni necklace she created.

"Did you write a whole bunch today, Mommy?" she asks.

"Yes, I wrote today baby," I say, and it is true. I filled pages of my journal with prayers and unanswered questions.

Over the next week, I attend online community meetings held by the ambassador to apprise embassy families of what we need to do during this time. These talks are speckled with encouragement to stock our cabinets in case we need to shelter in place, and we are informed of the best hours to go grocery shopping. At the store, I grab dried beans from the shelves, stock up on water and rice. At home, Evelyn and I play games and read books in the overstuffed chair that overlooks the valley from our window.

One afternoon, a storm rolls into the valley and we watch as the dark clouds dump sheets of rain and lightning bolts break the surface of the sky. The street below us is bathed in light, and I'm struck by how strange it is. If we weren't watching right now, the only notion of danger would be the low rumbling of far-off thunder.

A few days later we are reading in the chair when my phone rings, a muffled buzz buzz buzz on the dining room table. I knock several books off the ottoman trying to extract myself from the cozy nest Evelyn and I had built.

I feel my brows crease when I see Rebecca's name on my phone.

"Rebecca?" I say, and it's as much a question as a greeting.

"The hill near our house—it's on fire," she says in a breathless voice. "Have you heard anything about this?"

I tell her no and walk to the back bedroom to look out the window that faces the direction of her house.

"I see smoke coming from your neighborhood. Have you called anyone yet?"

"Yes, I reported it. But I don't know if it is safe to stay here."

"If it spreads it might not be safe to stay here either," I say, and turn to see Evelyn in the doorway. I turn back to the phone. "Let me call Ryan, see if he has a bead on it."

"Mommy, what's that?" Evelyn asks as she sees the air above the embassy housing darkening with smoke. I bend down on one knee and tell her to grab her favorite buddy from her bedroom.

"I'm going to call Daddy," I say. "You can sit all cozy right here and look at some books for now." When she returns with a buddy I wrap her in a blanket and put her in the chair in my bedroom. Ryan says he knows of the unfolding situation. If it gets worse, they will need to evacuate the families living over there.

"Where will they go?" I ask.

"They'll have to come to the embassy," he says. I ask if Rebecca and her family can come here. He tells me he should know more in a little while. For now, he says everyone should stay put.

"But check your go bags, all right?" he asks, and I confirm.

Our go bags are always packed in case of emergency. I don't want to think we are at the place of "in case," but maybe we are. I drag ours into the bedroom and heft the green hiking backpack onto the bed. I run through the contents: enough water for a few days, clothing and extra shoes for Evelyn and me, our important documents. I head to the kitchen to add some snacks and food supplies. With the bag fully stocked, I turn and sit on the bed and wriggle on the straps. It is heavy, but once I pull the support strap and heft it onto my hips the weight is manageable.

"Come here, Evelyn," I say and tell her we're going to play a little game. "Like Mommy Boat, but you are going to hang on the front of me."

"Like a koala!" she says.

"Yes, just like a baby koala."

She climbs onto the bed and then onto me. She holds onto me with her legs, and I see if I can make some laps around the apartment carrying her and the bag. *This is ridiculous,* I think. *I am most likely overreacting, but I need to know. I need to know that if I have to carry her and our stuff, I can do it.* And I can. I kiss her and place her back on the bed.

"Great! We did it!" I high five her and leave the go bag beside our bed.

The next twenty minutes are filled with restless silence as the darkening smoke cloud grows and sirens blare through the streets. Ryan calls with an update. They are evacuating homes from the fire. It is not headed our way and should be contained soon, but if Rebecca and the kids want to come to us, they are welcome.

Within the hour they are at our door. Her kids spill in with their backpacks and pillows, stuffed animals, and books. One of them is carrying a giant bag of cereal. I welcome them and we set the table with Play-Doh and other activities. Rebecca and I eyeball one another, and I close my bedroom door, the one with a window looking out toward the fire. I close our windows too, to shut out the whiff of burning. I pop up a kid-sized table for the little ones.

"Does everyone like macaroni?"

"Yes!" all the kids shout. Soon they are busy with a meal. Afterward, we place the little ones in front of a show, and the older kids go into the guest bedroom to read.

Rebecca and I are left on the couch together. I turn to her and we sit in silence for a moment, grabbing hands. We put our heads down and pray. The words don't seem big enough, but we say them anyway, "May we be surrounded by your love and protection. May we know you are good, You are here with us."

We make it through the rest of the day by doing a yoga video and reading books to the kids. We bake two cakes and have a merry unbirthday party, for lack of a birthday to celebrate. We play dance tunes to drown out the sound of helicopters flying back and forth to the fire with buckets of water.

For several days, Ryan stays day and night at the embassy. Rebecca and her family stay with Evelyn and me. I am grateful to not be alone.

When Ryan does return, I see he is tired but invigorated. I can't help but feel proud that he is who he is. In this troubling time, he stepped into his excellence. I think about his deployments, and how those were also opportunities for him to live into his talents. I had not looked at our time apart like that before.

We learn that the fire was caused by barrels that were ignited by more civil unrest that turned violent and destructive. Rebecca and Joshua are cleared to go to their home and assess the damage to their items, leaving the kids with us. Ryan and I put on a movie and pop several bowls of popcorn. The kids crowd together in their jammies, and Ryan and I sit at the table holding hands, waiting to hear from Rebecca.

Thankfully, most of their items survived. The house will need structural help as well as deep cleaning, so they are being moved to a hotel for the foreseeable future. The next morning they pack to leave. The kids enter the elevator as we adults pause to hug one another in the hallway.

That night Ryan and I lie on our bed, speaking in quiet tones of what it was like for him at the embassy. I show him pictures of the cakes we made.

"Is there any of that left?" he asks.

"Are you kidding?" I laugh. "Those kids licked the plates clean."

He grabs his phone and flicks through photos of families spread out in the hallways of the embassy. He tells me how the kitchen staff kicked into gear to feed everyone.

"Should I have been there to help?" I ask.

"You were helping here," he says.

I remember something author Lara Casey once asked during an online event for military spouses. I share it with Ryan: "Who will you be in the middle of the hurricane?"

"Were you who you thought you would be?" he asks, and I lean back onto our pillows. "Yes, yes, I think I was," I say.

"I think I was too," he says. I bring his hand to my lips.

"I'm going to write for a little bit before bed," I tell him, and I reach across the nightstand to grab my journal and pen. I reflect on the past few weeks and look ahead to moving back to the continental US soon—after spending nearly a decade away. We made it through this latest trial. I am both shaken and sturdy. I write:

> *I am still surprised we had fun and made memories regardless of the heaviness surrounding us. Life persevered even during a crisis, and I believe that in our home and in our community we were patient, kind, loving, and calm.*
>
> *I am holding on to the good times that presented themselves, and the ones we created from a thin air of hope. I did not hide my anxiety, but I did manage it by breathing throughout the day. Putting on yoga videos. Dancing in the living room. Pausing to let the sunshine hit our faces. Ordering a pizza dinner that flickered in candlelight. Reading delicious books out loud.*
>
> *We slowed everything down and learned to take it not even a day, but a moment at a time. Life can persevere, and love can break through my fear.*

I start to set my journal aside, but then I crack it back open again and scribble:

> *Upcoming move to Virginia: Who will I be?*

Ten days later I am pacing the kitchen. I cross the house and peer into the office. We have three and a half months before we leave Ecuador. Toys are scattered across the floor, books spill off the bookshelf, and a pair of kid-safe scissors lies on top of shredded scraps of paper. I sigh. There is so much to do before we go.

I cross back over to the kitchen and avert my eyes from the pregnancy test resting on a paper towel. Two minutes to go, and I am walking out the timer. I go to the window and look out across the valley, breathing in the sight of the immovable Andes mountains that stand soft in the landscape but strong in the distance.

Wherever we live, I'll miss this view. I will miss a great many things about this chapter of our lives, and yet I am excited about the next one.

I can hold the tension of both feelings. I know the path ahead for our family will not be linear. I cannot expect it to be. I cannot expect to have all the answers, even if I might wish to. The timer dings and I walk to look at the test.

It has a plus sign, with the Spanish word *embarazada* next to it, in case I had doubts.

I lean down on the counter, elbows on the cold marble and head bowed.

"Thank you," I whisper. "Oh, thank you."

I cross my arms over my abdomen and hug this burgeoning life. New adventures are starting, in more ways than one.

Who will I be? I ask myself again.

The Long Winter

THE CHANDELIER IN MY MOTHER-IN-LAW'S guest bedroom offers a shimmering glow. I stare at it and will myself to leave the cocoon of soft sheets, pillows, and blankets. When I lift my head, I can hear Evelyn's feet running back and forth to her Papaw, and I squeeze my eyes shut again. I could blame these long naps on being in this pregnancy's first trimester, but I think they also have an edge of depression.

Whatever relief and excitement I felt months ago, I didn't expect to be where I am now, lying in my in-laws' guest bedroom in Ohio, pumped full of the extra progesterone that is supposed to help my body hold in this new baby.

Three weeks ago, at the doctor's office in Quito, I saw the baby, a tiny gummy bear, its heart flickering on the monitor. We had confirmed the pregnancy through the embassy clinic, but following some breakthrough bleeding, we'd scheduled an appointment with a specialist in the city to see what might be happening in my body.

"The baby is fine," the doctor said, pointing at the screen and shifting in her tall black stilettos. I nodded and bit my quivering lip.

She traced a secondary shape with a long manicured fingernail. "But I do see a blood clot, in utero, no?"

I squeezed Ryan's hand and he asked if this was dangerous.

"Oh, no, most likely not," said the doctor. She said the bleeding might continue, and that she was prescribing a progesterone supplement. My activity is also limited. No intercourse, no intense workouts, no lifting anything heavier than ten pounds.

I think of Evelyn, at home, and how I will need to break the habit of sweeping her up into my arms. I wish I'd known the last time I would get to do that before this happened.

The doctor printed out gummy bear photos and her diagnosis for us, put it in a folder, and sent us on our way. My legs shook getting into the car, and I rubbed circles over my belly with my hand as I buckled my seat belt.

The doctor had provided some comfort to me, but she also gave me advice I didn't know how to follow.

"You, mama, you cannot keep worrying," she said, shaking her head and looking directly at me. "You cannot."

Not worry? She said it so casually, as if it were as easy as kicking off those high heels of hers and sinking into a pair of sneakers. My own nubby nails held the file tightly on the way home, gripping the papers. *How could I quiet my deep concern about the little one growing within?*

What does a parent do but suppress worry daily, just to survive and embrace the gift of these human beings in our care?

That night, Ryan and I talked through our options. His mom and dad were visiting us for the Marine Corps Birthday Ball and would fly back to Ohio the first week of December. Our family wasn't scheduled to move until the second week of January.

"I think you and Evelyn should go home with Mom and Dad," Ryan said.

I shrank back into myself on the couch. "But then we'll miss Christmas with you," I said.

"Yes. But you can begin to sort out our life back in America, and ... " he trailed off.

I looked at him. Maybe he was right. The fire and the riots had been an intense scenario. Ryan didn't want us here if something kicked up again.

"I do better when I know you're safe," he said, and so we logged into the airline to buy a flight for Evelyn and me.

Now Evelyn and I are in Ohio, and I am fighting to be content with our circumstances. I miss Ryan so much, and we still haven't told Evelyn about the pregnancy. We don't want to add this layer to her transition back to America. She is adapting well enough, with family all around her. I find myself in a daily state of overwhelm, trying to manage the burden of choosing to be away from Ryan for the holidays, the intensity of American highways and superstores, and decisions that need to be made at every turn.

What's wrong with me? There's a part of me that doesn't want to celebrate the baby at all. I don't want to get excited about the possibility of this new life. Maybe if I can contain my excitement, I could also contain my future grief if something goes wrong.

I'm holding everything in, and the effort pushes my body down deeper into the mattress, begging for sleep to take me back to a place where I'm not thinking. Back to where there is a baby and no bleeding. I bury my face in the pillows as I hear a "ping" from my phone.

I swipe to see a text from Katie. "Can I call you?"

Katie is one of my oldest friends. She was the first to hear about Evelyn's existence. She knows the ins and outs of our family and of me. She came to see us in Hawaii, Morocco, and Ecuador. On each trip, we forged a new set of memories.

When I have to leave friends at different locations, some of those friendships are for a time, and then they come to an end. But there are a special few who will always be with me. Katie is an anchor, a constant, and for this, I am grateful. She knows we're wading through a challenging time after the abrupt move from Ecuador and with this pregnancy.

I respond with a "Yes, call me!" and her name pops up. I swipe and put her on speakerphone, propping the phone on my pillow.

"Hi, you," she says. "I've been looking at flights."

"What?" I say, propping myself up on an elbow.

"Yeah. You are so close now, that's one of the things that's great about being back! Would it be okay if I came to visit for a few days after Christmas?"

I cover my hand with my mouth, laughing and also catching a lump in my throat. "I would love that. I would absolutely love it."

"Okay. I'll text you the details of my flight, okay?"

"Okay." I pause and swallow, "Katie? Thank you."

"I'm happy to do it," she says.

I pull myself from the bed and make it through the rest of the day. Over the next week, the excitement of Katie's upcoming visit helps me focus. I get to work. I choose the apartment we will move into in Virginia, working with a real estate agent to secure our best option. It still feels like a gamble because I'm signing an agreement to rent it sight unseen. I buy a bed for Evelyn, two vehicles, and a dining room table. I make an appointment with an ob-gyn in Virginia. I'll be twelve weeks along when we get there.

We make Christmas cookies and I take Evelyn into town to explore the library. We max out the number of books we're allowed to check out on the card my mother-in-law has lent me. I'm putting one foot in front of the other. We're out until dusk. Christmas lights start to twinkle outside the houses we pass as we navigate back.

After we unload our books, I pause by the kitchen counter and take out a pen and a piece of yellow scrap paper.

"Mama, come read!" Evelyn cries from the couch, pulling a pile of books around her.

"Hang on, baby," I call, and for the first time in months, I feel like writing. I scribble a poem:

She edged the van onto the street,
The homes lit up with lights, giant gingerbread

Sweet on the eyes
A flurry of snowflakes
Spattered across the windshield like crushed diamonds
Her daughter sang,
"Have yourself a merry little Christmas"
With a wiggle in her car seat
Dull clap of her blue mittens
And the mama looked at the same sights
But her heart was dull
"Where is it? When will I feel it?"
She prayed for magic
A backseat squeal made her hit the brakes
"Mommy, my favorite—look!"
A giant inflatable Frosty, tilted over to the side
Her daughter's laughter rang like sleigh bells
And then she heard her laugh before she felt it
Stiff at first, a throaty rumble
A burst of sound
And for the first time in weeks, she thought:
It may not be Christmas
Without him
But maybe I can still find Merry.

When I put the pen down I smile. The inked lines mean something. I captured a moment. I'm still in here, somewhere. Evelyn and I cuddle up under a blanket and I read aloud until my voice cracks. Snow starts to fall outside the window, and I lay my head back as Evelyn places hers on my lap. I wonder what it will be like to hold two children here.

The snow picks up outside, a flurry now. At first, it melts as soon as single flakes touch the ground. But I notice that when they group together, the snowflakes stick. They start to form their cold blanket across the earth, and as Evelyn lays her head next to her sibling's growing form within me, I try to let a hope, fragile as it is, exist in my heart.

Christmas comes and goes with warmth and kindness, and I am present but also removed. I build a dollhouse with my father-in-law, working together into the wee hours of Christmas morning. I hug my girl as she flies around the house with untethered joy. I experience flashes of this goodness from my child, my family, and our future move, but there's also a persistent veil of darkness.

This is a different feeling from the anxiety or depression I've encountered in the past. This is like my light has gone dim, like a candle about to go out. It's not out yet. Some days it feels like I'm protecting it, feeding it carefully, wanting to stay alight. Other days, I want to sink back under my covers, blow it out, and release myself like a wisp of smoke to the dark.

When Katie arrives, we book a hotel for a night away. We stroll through a quaint restaurant and shopping area. I am grateful she is here, but there is a part of me that wants to hide. I am not a confident, bright, and shiny version of myself. I'm behind a screen, looking out. She doesn't ask me to be anything I'm not. She listens when I feel like talking, and we walk. When I choose silence, she's next to me anyway.

As the sun sets, we have time to pop into a bookstore before our dinner reservation. I move toward the children's department, admiring the glossy covers and tantalizing titles on the shelves. I am pulled toward a section of board books. Above the books is a shelf of stuffed animals. I remember when I was pregnant with Evelyn and my mom visited me in Hawaii. I was worried then, too. We had gone to the Disney Store for a gift for a friend. In the checkout line, I saw a little stuffed Nemo. In the movie, the brave clownfish survives a trip across the ocean. Both he and his dad discover the value of being a family, but also the importance of letting go of fear in order to grow. I remember I held the plush fish in my hands and gently rubbed its fin.

"You should get that," Mom said. "It's okay to want to celebrate this baby." I knew she was right, but I hesitated anyway.

"What if—?" I asked that day. But she stopped me and we made our way to a corner of the store. I told her about my friends, their miscarriages, their fertility struggles. How could I dare for this to work out? How could I hope when I knew all that could happen?

"How could you not?" she asked.

I took Nemo to the counter and purchased it. Evelyn never loved that little Nemo like I did, but then, it wasn't truly for her. It was for me, a small talisman to hold the place of the baby to come.

Now, shopping with Katie, I pick up a stuffed llama and stroke its fluffy white body, look into its sweet little face, and imagine our new baby grasping for it. I need to start celebrating this life too. I tuck the llama under my arm and draw my eyes back to the bookshelf. I search for the author Nancy Tillman, one of my favorites, and select a book entitled *You're Here for a Reason.* As I read through it, I nod and sigh. It includes the line, "A piece of the world that is precious and dear would surely be missing if you weren't here." I know this is the one I need.

I find Katie and let her know I'm ready. She eyes the treasures in my hands.

"I'd love to get those for you," she says.

I hug her. "Thank you. But I want to do this. I need to do this."

She smiles and hugs me back. "I think that's wonderful."

On New Year's Eve, back in the guest room, Katie and I lie in bed, staring up at the chandelier together. We've spent the evening batting balloons around the living room with Evelyn and counting down to the new year at an early eight o'clock. Thank goodness four-year-olds can't tell time. We call it an early night.

"Do you think I'll feel happy again?" I whisper. And then I cover my face with my hands. I tell her I'm so glad she's here, but I just don't feel like myself. I keep thinking I can will myself to feel better. Instead, I get a glimpse of the me I recognize in wavetops, riding up and down on an ever-changing tide but shrouded in gloom. I keep reaching for an end to this pain, to the worry, to the wanting, but maybe I have to ride it out. Katie reaches to squeeze my hand, and I let her. I take enough deep breaths that I finally fall asleep.

In a week, Katie is back on a plane to New York City, and my in-laws pack up our two new cars, plus one of theirs, to caravan down to Virginia with me and Evelyn. We'll get Ryan from the airport in five days. The roads wind through American forests and fields, and I hum along to songs on the radio. I have spent much of December having to ask for help. I've needed naps, extra doctor visits, and support in giving Evelyn the love she needs as I muddle through.

Snow falls as we drive through the mountains of West Virginia. One flake, and then many, collecting on the glass, piling up together at the base of the windshield. I received what I needed by coming back to America early. Even so, I might feel this way for a while. I look in the rearview mirror and catch Evelyn sleeping. I place a hand on my womb.

Maybe hope persists, and it can grow—even in the dark.

Pandemic Blur

I HOLD RYAN'S HAND AS WE WALK through Walmart in Suffolk, Virginia, sometime in the middle of March 2020. Our long-awaited storage shipment had arrived the previous day. After days upon days of calling and emailing the shipment provider, we discovered that while our paperwork was received in December, no one in that office had pushed it through. Since January, we've existed with the new dining room table, Evelyn's bed, a pull-out couch, cooking utensils, and whatever we brought in our suitcases. Then, while the shipment was being transported, they lost track of it, which added another delay. Now it has finally come in, and not a moment too soon, as the world is starting to shut down.

Ryan steers the shopping cart with one hand, and Evelyn rides in the main basket, singing and playing with a dolly. It has a purple dress and long brunette hair, and Evelyn is twisting the doll on her lap into different dance poses—splits, side splits, a sweeping leap, and a cartwheel.

When I picked up Ryan from the airport in January, I hugged him and thought, *We're whole again. This is where it gets better.*

And in some ways life has. Our transition to Virginia has been both good and rocky. Like every move. Within a month of Ryan's arrival, he had to leave for a work trip. His absence pushed me to a Mothers of Preschoolers meeting in an attempt to make connections and friends. I added several women's numbers to the contact list in my phone in case of emergency—and hopefully for potential playdates. I talked to our neighbor, a kind man in the navy, who helped me open our manual garage door on trash day before Ryan got here. It's humbling to ask for help for such a small task, but I'm still restricted from lifting anything over ten pounds.

We surprised Evelyn with the news of the new baby with a "big sister" shirt featuring Elsa from *Frozen* and a stack of books about new siblings. Ever since, she's been walking by my belly and whispering, "I love you." She has also rolled around our bedroom floor in a fury, asking if I will still love her as much when the baby comes.

I tell her there's abundant room in my heart for two. And through her tears, I stroke her hair and tell her she can be excited and scared at the same time. I don't tell her I am both of these emotions, too, all day long.

Once my in-laws had helped us get set up, my parents tagged in, and during their visit, we revealed the baby's gender. A single blue balloon popped out of the box, surprising everyone, especially me.

Each new expansion in my body, each new bit of information helps. And the relief of being taken off the progesterone prescription at the sixteen-week mark is nice too. Still, I don't feel assured. I live my days counting down till July, and when anything feels off, any new twinge in my abdomen or the still-frequent smears of blood that appear when I wipe, I fall into a fog of terror. I feel like I can't enjoy this pregnancy. I am marking the days, getting through each one. It feels like waiting, not living.

The lingering darkness in my mind is always present at the edges. One gray evening I finally ask Ryan to sit with me.

"I think I need to go back to counseling," I say, my hands twisted in my lap, clenching and unclenching. He asks why and I try to explain that when I wake up every day I am ready to make it a good one. But then I feel pulled under. And when any small thing disrupts me, causes conflict, or feels intense, I fly off the handle. I cry or bubble over in rage, sometimes both, and all with Evelyn in view. It doesn't feel safe for me to navigate all these feelings alone.

"What feelings?" he asks me, and again, I try to unpack what it's like to be in a new place, again—pushing ourselves out to meet people, again. I'm going through the motions of getting us settled while not having any of the gusto I had in previous moves.

"I used to be able to muster the energy," I said, hands now tight together in my lap. "But I am tired. And I'm still scared."

I did not want to have to say it. I am admitting I'm not well. I feel like a failure for needing extra help again, for bemoaning any part of this transition. We made it back to America. I'm pregnant. I have a beautiful daughter and a husband who isn't deploying. Why can't I just be happy with that? But maybe this conversation is also admitting that I want to get better. My body is here every day, but my mind is frequently doing somersaults, dizzy, and upside down.

"I need more tools to bear the weight of this," I said.

"Okay," he responded. "I'm proud of you for taking steps to get what you need." He placed his hands on mine then, pried them apart, and kissed them. I booked an appointment for the next week, and I began to see a counselor online.

The counselor is also a military spouse. She is patient and kind, and she lets me release all my thoughts and emotions while guiding me to support myself. Throughout our sessions, we practice breathing techniques. She helps me name the hardships of returning to America. We talk about how I had wanted this relocation back to our home country to be easy. I thought I had waited long enough to be pregnant, so I hadn't anticipated complications. We were finally close enough to drive to family and friends, but the COVID-19 pandemic put an end to that with the shelter in place. I wanted so many things, and I wanted them my way.

I tell her I'm hiding from God, mad that this has felt hard. And she helps me develop a prayer. I order a green stone bracelet I wear on my left wrist, pausing throughout the day when my anxiety starts to rise from my heart to my throat. "He is sovereign. I am safe. He protects and provides." I say it at our bathroom counter, staring at my growing belly. I say it in the kitchen over a pan of scrambled eggs. I say it, and I say it, and I work to believe it.

I twirl the bracelet now, as Ryan winds the cart through the camping aisle, and we place a green gas camp stove next to Evelyn. The store is packed with people but it is hushed. It feels like everyone is darting back and forth between the aisles while avoiding body or eye contact.

The toilet paper is long gone, so we pick up paper towels and napkins instead. I don't see what else we buy, I'm so busy watching Evelyn in the cart, with the dolly doing cartwheels on her lap. She spins in a furious dance, hair flying, dress swooshing. I wonder if I should wash the doll when we get home. Another cartwheel, a massive spin in the air. The doll's motions feel like my brain, my heart, twisted and wild. We load everything into our van. It is the last trip I will make to a store for three months.

One day near the end of March, I'm sitting at the computer, staring at the screen so long that the text and graphics begin to blur. Like these long days at home. It is now week three of social distancing and quarantine due to the global COVID-19 pandemic.

We wake, eat together, clean up, read books to Evelyn, play with Evelyn, take turns working in separate intervals in our master bedroom turned home office, and repeat. I squeeze my eyes shut and reopen them, urging myself to focus as I scan over the last pages of this "special edition" newsletter I'm putting together for the embassy in Ecuador. Even though we have left the country, my position as editor was renewed, so I've kept working virtually. It feels odd to write encouraging bulletins, make craft lists, and offer creative workout ideas. I have one creative toe still in the community there, but I'm not physically present.

The one thing I leave the house for is ob-gyn appointments. My previous pregnancy, I went to all my prenatal appointments alone due to the deployment. Now Ryan is here, but I'm alone again. The irony is not lost on me. There are so many of these tiny losses that I can't count them all. Across the news and social media we see declarations that "We're in this together!" But I wonder, *are we?*

Out my bedroom window, I see our white picket fence. Evelyn is down below, playing with chalk and making mud cakes while Ryan circles the backyard, trimming up bushes. Our friends who are still in Ecuador aren't allowed out past their backyard fences, not even on neighborhood walks. I remember our sixth-story apartment there and wonder if we would be allowed to enjoy the fresh air in the community rooftop garden. If not, we would have been indoors for this past month without ever stepping outside.

In New York, Katie witnesses a constant stream of medical vehicles flowing through otherwise empty streets. My sister, Bethany, in Texas, is managing four kids at home while guiding her fifth-grade class through lessons over a screen.

The whole world is experiencing something as a collective, but each view is very different.

In the first fourteen days, I had energy and creativity. *We have done hard things before! We can do this! We're a military family! Gearing up and accomplishing the mission is where we flourish!* But as the days drag on, we're all starting to lose context. I'm losing exclamation points. We're wondering what will come next.

Rubbing my temples, I click to attach the final version of the newsletter to an email to my boss and hit send. I push back from the desk and head downstairs. Ryan is sliding the porch door open with one hand while carrying Evelyn in to wash her hands. I slump down onto our gray sofa and watch him. Ryan's presence in our everyday routine is also new—a big gain instead of another tiny loss.

This is the longest span of time we've spent with Ryan in person that I can remember. While I watch him, I know I don't just love him. I like him too. We've had to be friends during this time, and while we always were, this time and our limitations have drawn us together in new ways.

After he washes up Evelyn, he asks if she'd like to play upstairs with a snack, and she says she would. I stay sacked out until he comes to sit with me.

"So the governor of Virginia released a new mandate," he says.

"Break it to me," I say, closing my eyes tight.

"We're on a stay-at-home order now until June tenth," he says.

"What does that mean?" I ask as I rub underneath my belly.

"Only essential trips out. Food. Medical."

"Can we walk?"

"We can, as long as we continue to socially distance ourselves."

I feel sick. And angry. I am exhausted too, not from anything physical but from deep emotional fatigue. I calculate the months ahead. We have to stay apart from the world until four weeks before our baby boy is due. Then for three months after his birth we'll still need to shelter ourselves somehow because as an infant, he'll be lacking many natural immunities. What will we do to get through April, May, June, July, August, September, and October? I feel the weight of the days, months, and moments ahead. I rub the green rocks on my wrist. He is sovereign, I am safe. *Am I safe?*

I hear Evelyn laughing upstairs at the book she is listening to. The Ramona Quimby series by Beverly Cleary has become a solid favorite. When I asked her why she likes Ramona so much, Evelyn smiled and said, "She reminds me what it's like to be a real kid."

Is this time hurting her? I wonder. Her kindergarten year is coming up, but will she be able to attend in a school building, a real classroom? I want so much more than this present state of the world for her. But this is what we have. *Is it enough?*

"I need to walk right now," I tell Ryan.

"That's good," he says. "By yourself?" He asks and I nod. I walk to the front door and he follows. He helps me slide my feet in my shoes and I stroke his fuzzy hair while he's bent down. He needs a haircut. I can't remember the last time it's been this long. He ties my shoelaces and I wonder at him, his presence, and how it steadies me to have him here.

As I walk in our empty neighborhood I spin my green stone bracelet again. God is sovereign. I am safe. He protects and provides. I wonder if he can hear me. I wonder if I'll have to deliver alone at the hospital. I wonder if Evelyn will get to come to meet her brother there. Who will be with Evelyn when we go to the hospital?

A neighbor comes out their door for their walk and waits until I pass their driveway before they come down, giving me space. I think about how no one reaches to touch my belly in public this time. I'm rarely in public. When I am, the people around me spread out. I turn back to the neighbor, trying to catch their eye and trying to decipher what they are thinking, because I know what I can't hide: I'm pregnant in a pandemic.

Some neighbors, when they see me, have called, "Congratulations," and that elates me—a moment of normalcy. Being pregnant right now is advertising that you need protection on steroids. The bump is a private bodyguard, clearing the way before me. I keep walking. A steady trudge, my belly stretching in front of me. My thoughts are quiet, broken only by flutter kicks from my passenger. This baby knows nothing of current affairs. But he must know the three voices he hears daily now. His dad, his sister, and me. That's something good, isn't it?

The stretch of sidewalk ahead of me leads to the library. This was one of the great joys of this neighborhood: the ability to walk to a library! I remember how excited we were then.

As I trudge down the sidewalk I think about what else has changed. I didn't have Ryan during the first pregnancy, and now I do. I consider how far I have come in asking for and receiving help. I notice how beautiful the redbud trees are in front of the library, and how the sky is so blue, and how I keep putting one foot in front of the other. I pause on the path and wrap my arms around myself. Tears blur my vision as I stand and watch the purple-red blossoms blowing in the wind.

I wipe my eyes and walk past the double doors of the library, now locked. It is dark inside. The books line the shelves, waiting to be held again. Waiting to be read, understood, enjoyed, and explored. Checking out these stories is on pause, but our story continues to unfold. I wind my way back through the neighborhood and up to our home. I twirl the bracelet on my wrist and step in, shutting the door behind me.

In the Room

IN MID-JULY I FEEL THE FIRST twinges of labor at five thirty in the morning. I roll over, sit up, and nudge Ryan.

"We need to get our shoes on. I need to walk."

He doesn't move at first but presses his lips together with a "harumph" and rolls over.

I nudge again. "This feels like the baby is coming."

He turns back toward me and opens his eyes, bringing me to focus. "All right," he says. "Yes."

On the way out of the house, we pass my dad, sitting at our dining room table. He raises his eyebrows at us, and I tell him something feels different this morning. It is surreal to see my parents walking around our house, filling our sofas and their coffee cups, laughing, and doing chores alongside us.

Dad assures us he's got Evelyn when she wakes up. After almost five months of being a family of three on our own, now we are two more, and perhaps another one will join us today.

It was no small feat for my parents to quarantine for two weeks before making the seventeen-plus-hour drive across the country to be here, but they did it. If they couldn't have come, I would have called on a new friend for help. When I met Jo at the Mothers of Preschoolers group there had been something about her that connected us from the start. I saw her only three or four times in person before everything shut down, but her kind blue eyes and considerate heart met me when I was feeling off-kilter and unsafe due to the high-risk pregnancy. She's continued to check in on me with texts and phone calls. In a time such as this, our friendship has bloomed, and I am thankful.

I find I have to keep asking what is true. I have had bouts in these past five months of desperate loneliness. We have had friends in every place and carved a life into each duty station. I had developed a pattern that worked for me. Put me out there. Address the needs of the other members of my family. Get a routine established. Find the places and people we love. Ask what I can do professionally. Root ourselves in change, uproot, and do it all over again. In this place, day upon day, we have needed to learn new patterns. How to stay put. How to slow down. How to choose to enjoy life even when it isn't anything we would have recognized a few months ago.

Our walk around the block is fruitful. I step up and down the curbs, holding both of Ryan's hands. He reminds me that if it is time, I need to get rest, too. I need to get out of this heat and hydrate. I relent, but I know this is the day. I feel it. I head back inside and go upstairs to bounce on my exercise ball. Mom comes in and rubs my back. She doesn't say much, but her eyes are alight with excitement. I feel excitement as well, but also an edge of the darkness. *What is true?*

I am afraid of being in the hospital during these days of the pandemic. The hospital feels like a place where coronavirus lives. I am afraid of having the mask pressed against my face as I labor. They have already told me they can't give me any oxygen during labor, because any and all oxygen machines are being saved to treat COVID-19 patients. I used supplemental oxygen during Evelyn's birth, and it felt like the thing that got me through. I'll need to prepare myself to have this baby without it.

My body ignores my mind's uncertainty, moving through the stages of labor quickly, until I'm downstairs, gripping our kitchen countertop and looking at Ryan.

"We need to leave," I tell him and he nods, looking at his watch. My contractions are coming quickly, with little time apart. Mom is there, pressing on my hips to ease the pain, and I see Dad peering over the rail from the second floor, a look of concern mixed with admiration. His baby having a baby.

I remember getting into the van and then the hospital elevator, but each moment is a flash, a torn scrap of memory. Fluorescent lights. A swab up my nose, the COVID-19 test before admission to the labor and delivery room. I remember being covered in sweat, my breathing heavy, and my legs shaking. Kind nurses offered me popsicles as I wandered around the room, my mask at my chin. Little drips of grape flavor never tasted so sweet.

When I felt the overwhelming physical sensation to push, I remember a wave of fear. I don't want to release this baby into the unknown, into an uncertain world in an unprecedented time. I kept him safe all these months, inside the house. Inside of me. Now what? How can I keep him safe once he is out in this world?

What is true, what is true, what is true?

I see him, arriving in this world, helped by a team of nurses because the doctor doesn't make it in time. Baby Hunter's body is chunky, his head topped by a mass of blonde hair. I see him breathe and cry and I hold him, yet I'm told later that I can't stop repeating a question:

"He's okay? Is he okay? Is he okay?"

Ryan reassured me. "You did it. He's here. He's okay."

But I didn't stop asking. Ryan had to put his hands on my shoulders, look me in the eyes, and insist: "*Lindsay*. He's *okay!*"

And the rest of our hospital stay is a blur. I wish I could remember more, but there is a place in my mind where I crept away and experienced that time in small glimpses. The kind nurse who bathed me, so gently, in the shower. The "rate your pain" face poster on the wall, and how funny it seemed that their mouths were showing. How long had it been since we had seen a person's mouth? I traced every inch of Hunter's body with my fingertips, watching the flutter of his eyelashes and his pursed lips while he slept. On the morning we went home from the hospital, I remember the deer we saw grazing in a wooded area across from the hospital parking lot. They seemed unbothered, free, and calm.

Ryan's paternity leave gives us a few precious months together, even as his office is slowly returning to in-office hours. I'm grateful to have this extended time to bond as an expanded family. The Mothers of Preschoolers group arranges a meal train. Women I've never met drop by our house with tacos, zucchini brownies, soups, and fresh bread. They leave them on the front mat, like an offering.

We see you, the piping hot dishes seem to say from under their tinfoil covers. As we wade through our early days as a new family of four, we eat and marvel at how a community rises around us, even though it is a community we don't really know yet and can't be with in person.

The days after Hunter comes home are a blur like the early pandemic days. By the fall of 2020, Ryan is back in the office full time. Soon I'm waving goodbye as he leaves for a temporary duty assignment. I've done separations before, but this time I have to consider the chance of exposure to the virus. Then what? How sick would we be? As the only parent at home, I can't afford to be sick.

I read a post on social media from a mother who says she's feeling a "smothered loneliness." I laugh wincingly, because that is how this feels. A house where I am outnumbered by the needs of two. A home where I am deeply loved by two. Both are true.

"T-H-R-E-E, sun and moon and stars make three!" Tonight, Evelyn belts this song at the top of her lungs. It's seven fifteen, and I'm so close to bedtime. She sings it again, louder, and I grit my teeth. I turn my face into the shower stream and try to focus on letting the warm water wash away this day's dose of spit-ups and crumbs.

I am a team of one for meal serving, butt and tear wiping, laughter-inducing, dishwashing, rug sweeping, book reading, homeschooling, coat zippering, and the other duties of raising an almost six-year-old and a four-month-old. The shower curtain flaps open and Evelyn pokes her face inside.

"T-H-R-E-E!" Her eyes gleam, and my smile is a grimace. I want to stop the noise.

"Mommy—listen!" She opens the curtain wider and inhales for her loudest rendition yet: "T-H-R-E-E!"

I wipe shampoo from my eyes and put my hand on the curtain to yank it closed.

"Mommy, brother, and me make three! We're a team when Daddy is gone, Mommy—we are!" She grins at this and bounces to the corner where Hunter is sitting in his baby rocker. She tickles his toes and moves on to sing about the number four.

I rinse my hair and bite the inside of my cheek. I step out and wrap myself in a towel. Still dripping, I gather them both in my arms. She shrieks and giggles as a drop of water lands in her eye.

I look at her and say, "Can you sing that for me one more time?"

Our months continue like this, wading through opportunities, choosing sometimes to open up and other times to shut down. And I am always in the room, the one parent present daily, finding new ways to pour out love from an empty vessel. I discussed with my counselor that I feel I have no more tools left, except to keep going. She tells me there's no context for what this time has asked from me, which is a strange comfort. I focus on showing up for our family and slowly healing my body from birth. I don't yet know how to heal my brain from it all, but one day, someday, I hope I will.

It is March 2021. I pull out a carton of eggs at the grocery store and step back. A woman wearing a silver sparkly mask grins at me. I can tell she is grinning by the crinkle of the corners of her eyes.

"Your kids are so cute," she says.

"Thank you so much!" I say, and she asks me their ages and tells me about her grandkids.

Then she says, "I can't believe it has been a year of this."

"I know," I reply.

She shakes her head, and we exchange another crinkly-eyed smile and move on. After I load the groceries in the van, I go to push the cart in line with the others at the cart return. A man is there and I pause to let him go first.

"No, you go," he says. I nod gratefully, balancing the baby on my hip and grabbing my big girl's hand to walk back to the van.

"We're coming up on a year," he says.

"Yes," I say, but it feels incomplete. I turn back and say, "I hope you have a good day."

He waves.

When I get home from our grocery trip I see them, the social media posts and online articles, telling stories just like this one. We're all processing this pandemic year. I scroll back through my writing from 2020 and find my initial battle cry to get through hard times. After the fire in Ecuador, I wrote in my journal: *Life can persevere, and love can break through my fear. Who will I be?*

Here we are a year later, and I wonder: *Am I who I hoped I would be?* I think I was less. I think I was more.

Last night Ryan and I watched *WandaVision* on our gray couch.

"It can't be all sorrow, can it?" The character Vision says to Wanda. "What is grief if not love persevering?"

I squeezed Ryan's hand and held it until we said good night.

Maybe I had it right this last year. Maybe I had it wrong. Maybe both. Maybe all of life is a continuous circle of both/and. Life does persevere. It perseveres because of love.

Who was I last year? I was a mom who lost her patience and also played. I was a spouse who felt thankful and annoyed. I was a woman who lost hope and fought to find it. I touched grief and joy. I reset over and over. I reckoned and I adapted and I forgave and I felt rage. And I know that I loved. I know that I love.

It would be months before I can talk through the experience of Hunter's birth in counseling. I can't remember how many months, but I know it took a long time. Even longer to write fully about it. My avoidance of the topic still hurts my brain. I want a reason for why I don't want to bring up this most momentous occasion. It is uncharacteristic. There is a shame that circles me like a vulture, ready for picking. I made it through the tough pregnancy. Hunter is here. There is nothing for me to keep on about. Replaying the hard of the past few months does nothing, does it?

I've always celebrated the stories of our lives. I've been the memory keeper in our home, capturing the daily and fantastic details. Why can't I bring Hunter's birth into focus? I want to tell him about it one day. But I have it locked inside me. I tell my counselor this one day in a video session, in a shaky voice, after I've kissed and cuddled Hunter down into his bed for an afternoon nap.

She tells me it's time to go through it.

I cover my face with my hands and tell her I don't want to do that. I'm already having a physical reaction to thinking about it all again. My shoulders are up to my ears, my stomach is knotted, and the tears are already clouding my vision. But I also want to work toward healing, and so we greet the darkness together.

I close my eyes and start to retell the story I remember. She pauses me, only briefly, and asks if I can find God in the room. I open one eye, feeling almost silly. I am skeptical, but she says to try it. And so I do.

In the process of retelling Hunter's birth story, I still feel panic and worry, the weight of a baby coming into a world I don't understand. But somewhere in my labor I recall hands on my back. One of them was Ryan's, but there is another I believe was there.

A steady, firm presence that didn't desert me but came to witness and champion. And I hear it, a strong voice in my mind, as Hunter blossomed into the nurse's hands:

"He's here, he's safe. And I trust him with you."

I open my eyes in wonder and relief. *Can I believe this?* There is a hidden critic within me who is already raising one eyebrow at the whole scenario. But I wrap my arms around that critic and hug her, willing her to believe. What is faith if not a gift that makes a way to move forward?

After the counseling session, I stay in our bedroom and breathe. I open my journal and dedicate a space to Hunter. I write to him, and this is what I have to celebrate.

> *Through a pandemic and a blood clot and an international move, you arrived into this world, and you made our lives better than "okay." It was not an easy year to have a baby. Laboring through a mask, keeping you and your sister home to protect your lack of immunity, missing the support that would normally come from in-person groups.*
>
> *And yet somehow, it was also a year like none other.*
>
> *Neighbors cheered on your growth from across the street. MOPs mamas we had never met dropped food off at our doorstep. Deep conversations about "how to get together" opened boundary doors (even painful ones) with family and friends. By the grace of your growing fingers and toes, I have come closer to our Creator. You brought me back to life in so many ways, baby.*
>
> *You are the exquisite gain in the loss of the last year. You are here. And we are okay.*

Rooted in Hope

I ADMIRE THE TREES in the parking lot as I walk into my appointment. A few have begun to welcome the warm orange tinge of autumn. The rest stand green and tall, hanging onto summer. Autumn in Virginia is lovely. I feel a pang of sadness that by this time next year, I'll be looking at a different landscape. We are not sure where we are moving, but in the four months we have left in 2022, I'm slowly starting to pack, organize, and pare down our items.

After I check in at the front desk, I don't have to wait long to be seen. Soon the x-ray technician folds my leg into different positions and slides the moving platform I'm lying on to get all the right angles. I marvel at how with a few clicks from inside her booth, she collects information about what is going on beneath my skin. Another nurse leads me back to my room, and I climb onto the exam table. The paper covering crinkles and sticks to the backs of my thighs. It strikes me as funny that my legs dangle off the table awkwardly, not reaching the floor, just as Hunter's legs did at his recent two-year-old appointment. Middle-aged, young, or old, we all need a check-up every once in a while.

The doctor knocks and enters, and I smile at him. We've met on and off for the past two years since my knee pain began, but this is the first time I've seen him smile without a mask. His smile is genuine, with kindness reaching up to the creases of his eyes.

"So, Lindsay," he says, sitting on the black leather stool and wheeling toward me. "I have news."

Immediately apprehensive, I ask, "What kind of news?"

"You have your surgery coming up for your left knee, yes?"

I nod and flex my left leg, the one with the brace around it. We are squeezing in a repair for my torn meniscus before our move.

"But today you are here about your right knee."

"Yes." I stick my right leg out as if to prove it still functions and suggest that maybe the niggling pain and inflammation I've been noticing is like a passing shoulder shrug. Nothing to see here, folks.

"You have arthritis in your right knee." He delivers the results with raised eyebrows.

I sigh and tell him I am not surprised. My family has a history of arthritis, and I was a professional dancer for years. In pointe shoes. In heels. On a cruise ship. I've been a cat onstage as well as a dancing frankfurter. We laugh about it, and he tells me he used to play professional soccer in Turkey. He pauses and looks at me.

"It's frustrating to not move the way we used to," he says. I gulp hard. I look down at the bench and scratch at the paper.

"You miss it, don't you?" he asks. I nod. This is turning into one part paid orthopedic consultation and two parts free therapy. "You do have options, you know," he says.

"Tell me all of them," I say earnestly.

"You have to strengthen your body," the doctor says and hands me a list of exercises that will help the muscles around my joints. "Arthritis does not go away, but you can lessen the pain. You can choose to act now and help future you," he says. I nod.

That night, I pull a pillow across Ryan's lap on the couch. I flop down on it and debrief him on the appointment. I cringe at that part where I admit I have to make time for exercise.

"It is no longer an option I can put off. I either start putting the work in now, or I imagine I start to slowly rust in place, like the Tin Man." I picture myself creaking along after our children. He looks at me and kisses my forehead. "You can do this."

I nod but I also clutch the pillow a little tighter and pour out my fear that I won't be able to keep up with the new plan. It is a time investment and a change I was not anticipating. I recognize that this evolution isn't a one-time deal. There is no check-the-box-and-I'm-better. This is a renewed health commitment I have to make and maintain. I moan about the hardship of the last few years.

Ryan looks at me again and says, "You're talking yourself out of it before you even begin. I hear you, but you know what you need to do."

I start making slow adjustments, and as I'm reading *The Comfort Book* by Matt Haig after a yoga session one morning, I stumble upon this quote by Tara Brach:

> *Perhaps the biggest tragedy in our lives is that freedom is possible, yet we can pass our years trapped in the same old patterns. ... We may want to love other people without holding back, to feel authentic, to breathe in the beauty around us, to dance and sing. Yet each day we listen to inner voices that keep our life small.*

The last two years have been characterized by the fog of pandemic life, shrouding me in questions over each decision I've made. There have been swaths of anger and fear. But there have also been moments of deep gratitude, joy, and goodness that provide a place to plant the roots of hope. When I talk with friends and family, we all have memories of celebrations and struggles that occurred in the last two years. We all have choreography we've been dancing. I need to check my rhythm. Are my movements in harmony with the present, or am I off step, listening for the music of the past?

~

The room grows dim as the winter sun fades early. I am lying flat as a pancake on our rental bed in a dark Airbnb. My body is cramped with fatigue. My eye is twitching. Ryan is out shoving items into a hastily rented storage unit. On December 6, 2022, we finally received orders for our next move: Texas. By that time, the government movers couldn't schedule our pack out from Virginia until January 6. But our lease says we need to be out of the house by December 31. Hence the storage unit. And the twitching.

Thank goodness I've been packing up the house over the past few months. The sound machine hisses through the monitor, and I say a prayer of gratitude that Hunter is already in his footie pajamas and sleeping soundly.

This is our seventh military move in fourteen years—our third move during the Christmas season—and I am over it. On this particular night, I am trying to remain ever so cheerful and hopeful for my two children. However, I am also bone-dragging weary and soul-dragging defeated. I am still recovering from knee surgery. It is better, but I am often stiff.

The bedroom door creaks and I look up. Evelyn is in the doorway in her fuzzy Minnie Mouse pajamas. She pulls a book from behind her back.

"I've got *Despereaux*, Mama!" she says and crosses the room to stand next to the bed. "And clean teeth!" She proves this by puffing the freshest breath in the land onto my face.

"Mmmm," I muster, rolling toward her.

"Just one chapter?" she asks, placing her hand on my shoulder.

I scootch over and open my arms to her. She crawls in, and we crack open the *The Tale of Despereaux*. We follow author Kate DiCamillo's tiny mouse hero into a dungeon. The guard, Gregory, lifts Despereaux in the palm of his hand. Gregory looks at Despereaux, and tells him, "You, mouse, can tell Gregory a story. Stories are light. Light is precious in a world so dark. Begin at the beginning. Tell Gregory a story. Make some light."

My voice catches in the back of my throat. I have adjusted,

transitioned, and changed from place to place so many times now that it should be second nature. I remember that a friend smiled at me the other day when I told her how we were moving yet again.

"You must be a professional by now!" she said.

I summoned the confidence to tell her, "I am, and I am not."

I know I have fallen in love with places and people, held them close for a time, and learned to let them go. I know which rooms of a home to unpack first and how to leap into forming deep friendships fast. I know remaining married after tough deployments can be a battle for even the surest hearts. I know I have been both lost and found.

Each phase of military life has tested me, has asked something new of me, and there have been times I couldn't see anything but darkness. Times where I had to fight to find the light, experience it, and allow it to shine back in. On this cold December night, I brace myself to do it again.

I squeeze my girl in closer as I continue to read the words on the page, shining like sunbeams into this tumultuous time. For a moment we are caught together in warm comfort, bolstered by characters who continue to seek the good and battle for a meaningful existence. These moments, to me, are everything.

In the morning, I wake to Ryan sleeping next to me in the bed. I can hear the first rustlings of Hunter and Evelyn waking in their bedrooms down the hall. I shake Ryan gently.

"I'll be back soon," I murmur in his ear.

He nods and says, "Go."

I pick up a coffee on my way back to our now-empty house. Our things are packed and gone, and today the cleaners are coming. I greet them at the front door with a big smile and gratitude for the ease their service is bringing to this part of our transition.

While they go upstairs, I sit down in the middle of our living room floor and pull out my journal. Each movement echoes off the bare walls. One of the cleaners comes down the stairs.

"Oh!" she says in surprise when she sees me. "You don't have to stay." She explains that the rental company is fine with me leaving her the keys to lock up. I confirm this via text and get ready to go.

Before I leave our home, I pause and look back. We weathered challenging times and loved one another in this house. I get in the van and start to drive away, but I take the long road. I pass the small neighborhood pond and laugh at memories. Evelyn named a pair of geese there Agatha and Pegasus. I slowly curve in front of the library, thanking it for the stories I admired from the outside, for the day it finally opened again, and for the stories that unfurled from every bountiful page.

I drive down to Bennett's Creek Park, rolling down the window to feel the breeze in my hair, reminiscent of all the times it blew behind me when I biked down this same path. I park and pull out my journal and pen and put them to use.

This next move doesn't feel like a finish line, but in some ways, it is. Ryan and I have decided Texas will be the end of our journey as an active-duty military family. The Longhorn State might not be our grand finale in the Swoboda family production of *Places to Live.* I have talked at length with other military friends and find there are two camps: those who know exactly the place they want to be when military life is wrapped and those who are still guessing where to be. We are in the guessing group, but I know who I want to be with, and for now, that is good enough.

It feels like the end of an era, but the impact of this time will live on, for better or worse, and we get to choose which. I can apply everything I have learned and still discover there is more to learn. *Isn't that humbling? Isn't that life?*

I close my notebook. It is time to move, and as I bow out of the park, I pray for all the unknown steps ahead. Our family is once again perched on the edge, looking at the beginning of a new stage.

I trust it will be filled with people and places that shape, stretch, and sharpen our good, brave tendencies while gently encouraging us to let go of what is not serving us. I know how precious it is to believe that there is good in the world and that there is good left to be discovered in my own story.

I will be no less and no more than how I have grown thus far. I will work to stay rooted in hope, for while change will always be challenging, it is also a constant. I can fight change or continue learning to embrace it.

A WORD FROM THE AUTHOR

Got to Go Through It

MY TWO-YEAR-OLD SON sprints down the hallway, grabs my hand, and starts pulling.

"Whoa, where are we going, buddy?" I ask.

"We go to the liiibarrry," he says.

"Oh, I don't think we're going to the library today." Concern creases my brow as I brace myself for a potential toddler meltdown. But he pulls me into our front room, jumps up on the couch, and looks expectantly at me. My heart bursts.

"Is this the library?" I ask, and he nods.

For the past few months, I've worked to fashion this room into a cozy office and at-home library. This mission has been accomplished. For Hunter, this is the "liiibarrry."

"Go pick a book then!" I say and watch as he combs our three bookcases for his favorites.

We open the well-worn pages of the Elephant & Piggie books, as well as *Llama Llama Red Pajama* and *Lyle, Lyle, Crocodile*. My son's warm body presses into mine as he listens. The stories bring us closer together. He asks for one more story before a snack, and when I nod he rushes over to the stack of board books on the bottom shelf. His eyes and fingers scan through the pile until he finds the perfect one:

We're Going on a Bear Hunt by Michael Rosen.

Oh, yes. We've got to go through it.

~

To the reader who is not connected to service life, thank you for reading our story. While these pages speak to our specific experience as a military family, I hope that you will still find glimmers of the familiar. Do you know a military family? Reach out. We're pretty great friends to have, and we might need your friendship.

To the reader who is connected with the service, thank you for reading. Not everyone is called to this military life, but if you were, or are, there is an opportunity in the abundance of changes and challenges to choose to celebrate, to find a purpose in every season.

As you go through your own story, here is an invitation to be present.

To forge relationships.

To ask your extended family to love you where you are and to accept what you can give.

To endeavor to do the same for them.

To see the magic of being on a mission together, as a family.

To realize there is strength in asking for help.

To find the joy of saying "yes" to new places and people and what they offer—and all you have to offer.

To learn there is beauty in holding on and also in letting go.

Onward,

Lindsay

Acknowledgments

AS A CHILD, I LOVED WRITING THANK YOU NOTES to our family, friends, teachers, and others of great influence after they bestowed a gift. Whether it was a gift of material, time, or belief in me—it was a special process to hold pen in hand, pause, and acknowledge their role in uplifting my heart. I found that when and where encouragement and support came, hope followed soon after.

It is an absolute honor to recognize as many as I can for helping this book find its feet. Thank you, thank you, thank you.

My gratitude goes out to my great and gracious God. I often think of Your perfect timing and provision. You gave me a gift of words, and I will continue to try to steward them well. I've worked to turn my heart, mind, soul, and trust toward you. Nothing is impossible with You.

Ryan: my spouse, my champion, my adventure companion, my friend. The older we get, the more I appreciate our love and what we fought for and are building together. I love you, and I love our life. Thank you for every moment you believed in me before I did, for every cup of coffee and snack you sat on my desk while I typed away. Because of you, this book made it to the finish line.

To my awesome kids, Evelyn and Hunter. You both shine a light in the dark. You bring encouragement with your unrivaled imaginations and relentless need for read-alouds. Being your mother is a gift and a constant act of creativity. I love how you help me stop, notice, and live more bravely. Thanks for believing I could go for this big dream, and celebrating it all along the way. From our favorite Nancy Tillman, "Wherever you go—my love will find you."

Our parents: Robert and Doreen Crabtree and Brad and Wendy Swoboda, gosh you know how to cheer a girl on. You've shown up to watch kids, cook meals, and arrange a mini-writing retreat, and pushed me out into quiet places when I needed the space to create. You love our kids so much, and I'm grateful for every ounce of your energy and time you shared to make this happen.

Thank you to my sisters, Bethany and Christina, for picking up panicked phone calls and getting me laughing again. For Kristin and Rachel, for your many deep conversations about life, growth, and faith, and for reading books together, no matter the distance. Katie, you've said, "You can do this," since day one. Thank you.

Thank you to Callie Feyen for encouraging me to write. What you helped me believe about art and story and continuing to craft propelled me to keep plugging away at the manuscript. No matter how long it took, you told me to keep writing to find out. I am grateful for The Exhale Creativity Group, created by the writers of Coffee + Crumbs; you gave me the permission and confidence I needed to come to the page.

Thank you to my *Legacy Magazine* ladies: Abi, Courtnie, and Jess. Your talent for uniting voices, art, photography, and writing is inspiring. You pushed The Work of Words Workshop forward, helped me cultivate skills as an editor, and brought a passion for storytelling and helping our service community, which I carry forth. For the Work of Words writers, thank you for trusting me with your stories and letting the words work!

The Elva Resa team was instrumental in building this book. Karen Pavlicin, your attention to detail, wisdom in the industry, and intentional direction helped me through each step. Alexander Pavlicin, your professional touch and pep talks have been much needed! To my editor, Terri Barnes, thank you. Your writing made me want to write our story, and your editorial eye helped me organize and dig deep for the necessary story pieces to tie together. Thank you to my mentor, Corie Weathers. Corie, I will always be grateful for your impact on my writing, our marriage, and the example you set to go forth with great courage. Lindsay Wilkins, your words and art have long inspired me. To have your work on the cover is everything I dreamed of. Thank you.

I am forever inspired by our local community. Wild and Free Mamas, you listen without judgment and share your gifts, and we cheer each other on. I love raising our families together. You feel like home to me. AHG women: Thank you for prayer, a safe place to land, and reaching forward with faith. Thank you to our dear co-op, which pours love and support onto our kids. Thank you for the good energy and delicious brews of MildFire Coffee, and thank you to local libraries for your quiet spaces to focus and write, as well as your book-filled shelves holding great reasons to get distracted.

I appreciate everyone from past and present who allowed me to use your names in this book. Your early feedback from the chapters I sent strengthened my spirit and helped me know this work was worth pursuing. Thank you for being part of our story.

To every United States Marine we've met along the way, you will always hold a special place in Ryan's and my life, especially the Marines while we were stationed in Morocco. I will never forget how you played, read to, and cheered on our Evelyn and the many laughs and meals we shared. Thank you to Mary Regner, who took the time to walk with me as a new spouse. Your patience, energy, and kindness have always stayed with me.

I am grateful for our entire military family. You are some of the best people we know, and you are willing to fight for one another. You are willing to show up in the hard places. It is no small thing that you don't give up, and I will always be proud that the military life has been part of our family's story.

I am grateful to have walked among the military spouse community. It is because of you that I wanted to share my story. In each stage, I found a military spouse friend. In each tough transition, some listened, dropped a meal, sent help, said hello, and pushed me on. Our small acts of showing up are not insignificant. We can do great things together. No one else has lived this experience, has picked up the many fragmented pieces of our lives, and supported their service member the way we have. You are not alone, and I am in awe of you. You are stronger and braver than you think, but please also ask for help. When you feel your hope is running out or your courage is wearing thin, reach out and find a friend.

Thank you to the dancers I have been fortunate to work with or instruct. I always shared the quote commonly attributed to choreographer Ulysses Dove with my dancers: "Dance because you love it. Dance because you live it. Dance because you have nothing to lose and everything to share." It is with this same hope that I think of you, dear reader. If you're not a dancer, add your own verb. Run because ... Write because ... Hope because ... whatever it is, keep showing up and sharing.